P9-CAH-839

AIDS

The New Workplace Issues

AMA

MANAGEMENT BRIEFING

Contents

rkplace issues.

AIDS, the new
88205512

Introduction

The AIDS epidemic has affected the workplace on two levels. First, there is the problem of an appropriate response to a person with AIDS. Second, there is what David B. Dunkle, medical director for Hershey Foods, calls *AIDS-induced panic syndrome*. In his message to company employees, he defined the term as the "frightening behavior resulting from the vast amount of misunderstanding about AIDS."

Although Dr. Dunkle's remarks were pointed toward the fear associated with casual contact (sharing typewriters, telephones, water fountains), he might well have noted tremors in other corporate areas. In the early 1980s, more than a few company lawyers balked at the idea that an "infectious" disease such as AIDS could be classified as a handicap. Health insurers, with their own form of AIDS-induced panic, fought for the right to screen policy applicants. A person with AIDS would, on average, submit medical claims totaling $140,000, according to estimates prepared by the Centers for Disease Control (CDC). The insurers passed these concerns on to corporate clients and, of course, to their lobbyists.

Like much of the misinformation about AIDS, the $140,000 medical cost estimate (actually, an extrapolation based on a small number of early cases) has yielded to more rational numbers. The lifetime medical costs now average about $75,000—lower on the West Coast.

Among other developments that have helped clear the air, especially as AIDS relates to employment issues, are the following:

1. *Employers that dealt with AIDS early on shared their experiences.* In 1985, companies such as Bank of America and Wells Fargo allowed the American Management Association to publish or circulate their corporate policies on AIDS. These provided valuable models for others. In 1986, a group of 15 San Francisco Bay Area employers pooled their experiences, resources, and wisdom to produce a manual, *AIDS in the Workplace.* This became part of the San Francisco AIDS Foundation's workplace education package, released later that year. Spokespersons from other firms (Levi Strauss, Pacific Bell) freely shared their experiences with the press and with other companies.
2. *Other employers mobilize.* In October of 1987 ("AIDS Awareness Month,") the Orange County Business Leadership Task Force on AIDS and Alcohol and Drug Abuse distributed its handbook, *Facilitating AIDS Education in the Work Environment,* to 10,000 California employers, with development funded by Pacific Mutual Life Insurance Company. In that same month, The Allstate Foundation summoned 150 corporate leaders to Chicago for the first of its public issues forums. Under the banner of "AIDS: Corporate America Responds," the Allstate attendees formed task forces to address such issues as education, policy formation, and philanthropy. Earlier in 1987, the National Leadership Coalition on AIDS announced formation of the first nationwide organization for mobilizing the private sector. Meanwhile, various business groups on health (especially in New York and Washington, D.C.) were working quietly to help corporate members establish policy.
3. *The legal community removed virtually all doubts about the classification of AIDS as a handicap.* The early spokespersons reasoned that the Vocational Rehabilitation Act of 1973 would apply. Subsequent decisions* supported this position. These,

***Local 1812, American Federation of Government Employees* v. *U.S. Department of State* (1987); *School Board of Nassau County, Florida* v. *Arline (1987)* and *Chalk* v. *The Orange County Board of Education,* 9th Circuit Ct. of Appeals, among others. See Chapter 6.

plus the extension of specific state and municipal statutes to include HIV infection as a disability, have reinforced the classification.

4. *The case management approach to treatment has gained credibility, facilitating high quality of care at costs far below conventional hospitalization.* Equitable Assurance and Aetna provided early leadership. Now, virtually all carriers offer some form of case management. The early experiences from corporate clients indicate that the programs can meet cost and quality of care objectives. (Briefly stated, case management uses a case worker, generally employed or contracted by the carrier, to work with the patient, family, and medical doctor to facilitate the best quality of care. Applicable to a variety of catastrophic or life-threatening illnesses, case management is said to be especially promising for AIDS, in that it opens the way for less costly home and hospice care).
5. *Educational material has flooded America.* Some of the material (San Francisco AIDS Foundation, American Red Cross) is targeted to the workplace.

The newspaper headlines underscore the change in climate. In 1985, the news read "Many firms fire AIDS victims, citing health risks to co-workers " (*The Wall Street Journal,* August 1985). In 1987, the story was "Area companies educate workers on AIDS" (Westchester/Gannett Newspapers, September 1987). Two years ago, insurance firms focused on the legal issues surrounding HIV testing. (Major insurers in Wisconsin galvanized against a state law protecting confidentiality; the trade associations for both health and life insurance drafted policy statements on the importance of testing applicants.) In 1987, three of the most important private-sector mobilizations were funded by the insurance industry (Allstate, for the Allstate Public Issues forums; Pacific Mutual for the Orange County handbook; The American Council of Life Insurance and the Health Insurance Association of America, initial funding for the National Leadership Coalition on AIDS and for the Red Cross "Working Beyond Fear" series).

MEDIA COVERAGE AND WORKPLACE REALITIES

The popular press produces stories weekly. The CDC continues to roll out its chilling statistics.

Still, the vast majority of American employers have yet to encounter a first case of AIDS. As of this writing, the CDC reports about 50,000 cases of AIDS in the United States. The incidence, far below that for automobile accidents, cardiovascular illness, or cancer, accounted for about one percent of the national health care costs in 1987. Of the 50,000 cases, only about half were employed at the time of their diagnosis, according to some estimates.

In short, the AIDS epidemic has been far more visible in the press and on television than in the workplace. As one of our respondents put it, "It's an epidemic without a face."

Moreover, many employers who had to deal with a person with AIDS found the problem far less onerous than anticipated. The human tragedy outweighed other concerns; a common sense approach, with support for the individual and appropriate education for co-workers, proved workable. Interviews conducted for our 1985 publication revealed diverse opinions on how a company might respond. Similar interviews conducted for this publication caught a recurring theme: "we will likely respond as we would to any other illness or disability."

A broader view reveals larger problems.

To date, the workplace issues have been narrowly focused. The legal community gave attention to the rights of the individual vis-a-vis a safe environment for co-workers; the insurance industry, to cost exposure. Attention so directed overlooks the estimated 1 to 1.5 million persons who may be infected with the human immunodeficiency virus (HIV) but show no signs of illness. An appropriate response from the employer's standpoint has yet to be formulated.

Also overlooked are those persons who must care for someone with AIDS at home. If the trend toward home care continues and case management approaches live up to their projected potential, more employees will face the drain on time and energy associated with such practices.

More important, the incidence of infection has not yet peaked.

The CDC estimates that 250,000 cases will be counted by 1991, with a proportional increase in medical costs.

The solutions, as they are now being proposed, focus on education, on insurance provisions that facilitate alternatives to hospital treatment, and on the development of in-house counseling and referral services. Basic to these approaches is what some of the company spokespersons interviewed for this publication call an "atmosphere of trust." The company's position (via education programs and policy formation, for example) creates a climate in which persons with HIV infection can seek appropriate medical consultation.

A proposition. As more people gain direct experience—a friend, a colleague, a business associate—panic will lessen. The whole workforce won't be infected. Casual contact won't spread the disease. Incipient panic should yield to sympathy and concern.

The following chapters address this situation as follows:

- Chapter 1 looks at how various companies have responded to AIDS-related workplace problems, with special focus on actions taken among major employers in the San Francisco Bay Area. The efficacy of this (or any other approach) might be judged by the extent to which it meets the needs of special employee groups—those who are at risk, as well as those who have tested positive for the HIV antibody or who care for a family member at home.
- Chapter 2 surveys the personnel issues—how managers have responded to rumors that someone has HIV infection, how employers have provided reasonable accommodation, and the like.
- Chapter 3 takes a close look at policy issues, drawing information from approximately 20 organizations that have initiated AIDS-specific policies.
- Chapter 4 examines the various educational resources now available.
- Chapter 5 provides specifics on the cost containment potential of a case management program.

- Chapter 6, contributed by Arthur Leonard of New York Law School, puts the current legal issues into perspective.

The publication is intended as a tool for corporate decision makers in establishing policy and working through the specifics of workplace education. No publication can substitute for firsthand consultation with knowledgeable individuals in the legal, medical, and adult education fields. Readers are advised to seek such consultation, as appropriate.

QUESTIONS ABOUT THE WORDING

Given the magnitude of the human tragedy, a quibble with words may seem inappropriate. The language used to describe the epidemic has, however, led to confusion. When the CDC wrote the first definition for acquired immune deficiency syndrome, it did so before the virus was identified. Thus, the emphasis was placed on the syndrome (the characteristic symptoms) rather than on the root conditions (a virus that damages the immune system).

Consequently, an individual did not, for purposes of classification, have AIDS until one or more of the characteristic symptoms appeared (Kaposi's sarcoma or *Pneumocystis carinii* pneumonia being the most common). The definition has been recently broadened to include (a) presence of the virus as shown on an HIV-antibody test and (b) any one of 19 opportunistic infections.

The term "ARC" (AIDS-related complex) describes a less severe condition, whose symptoms include loss of appetite, weight loss, fever, night sweats, skin rashes, diarrhea, fatigue, lack of resistance to infection, or swollen lymph nodes. Positive results on an HIV-antibody test are also necessary for a diagnosis.

The terms "AIDS" and "ARC" cause few problems. Not so with other words:

"AIDS Test" and "AIDS Virus"

The first of these terms has been subject to challenge because it implies that persons testing positive have AIDS. Moreover, the test measures antibodies for the virus, not the presence of the virus itself. Thus, the term "HIV-antibody test" is preferred.

The term "AIDS virus" is subject to a similar challenge, in that it equates presence of the virus with the full AIDS syndrome. Again, it is believed that only about half the persons with HIV infection will develop symptoms.

"Exposed" to the Virus

One is exposed to a cold by being in the same room as a person with a runny nose. Popular use of the term suggests casual contact—shaking hands or sharing personal belongings. Unfortunately, many people (including reporters) have used the term to talk about a virus that is transmitted by blood or semen.

It seems more precise to talk about people who have "contracted the virus," a condition that can be detected with an "HIV-antibody test."

"Infectious" v. "Contagious"

Although the dictionary definitions show little or no difference in meaning, many spokespersons prefer "infectious" when talking about HIV. Such people feel the popular usage associates "contagious" with a virus that can be transmitted by casual contact.

"There is No Known Evidence..." That the Sun Will Rise Tomorrow

This is probabilistic language—the carefully chosen words of a scientist schooled in empirical methods of observation. To the scientist, the phrase indicates virtual certainty. It's quite proper to say that there is "no known evidence" that any event of nature will repeat—only probabilities.

Unfortunately, for many years government scientists indicated that there was "no known evidence that the immunodeficiency virus could be transmitted by casual contact." Small comfort to those who were not attuned to the vocabulary of the laboratory. For the man on the street, the phrase suggested that the jury was still out, that not all the evidence was in, and that the whole interpretation would likely change. ("If they don't know, who does?") Much of the epidemic of irrationality is still fueled by the feeling, "They just don't know enough about the virus."

Only recently has the CDC changed its approach, stating positively that the virus cannot be transmitted by casual contact.

1

How Companies Have Responded to AIDS-Related Issues

Companies in the San Francisco Bay Area became the first corporations to respond, in an organized manner, to the employment issues surrounding AIDS. They also became the first to share their experiences with other employers.

At Bank of America, Nancy Merritt, program manager for employee relations, led a five-member task force toward formation of corporate policy and establishment of support. The policy, implemented companywide in 1985, recognizes that persons with "life-threatening illnesses including . . . cancer, heart disease, and AIDS" may wish to continue in normal pursuits, including work. As long as such persons can meet performance criteria and their conditions do not pose a threat to themselves or others, they will be treated consistently with other employees, according to the policy.

At Wells Fargo, Bryan Lawton, director of employee assistance programs, launched a companywide training program, using a carefully worded "before session/after session" questionnaire to gauge the impact. Had the attendees changed their attitudes about, say, transmission via casual contact?

At Pacific Bell, medical director Michael Eriksen assured employees that "sharing telephones and breathing the same air pose no risk to co-workers." Although the company did not modify its policy handbook to include AIDS-specific wording, continued employment for persons with HIV infection became de facto policy.

At Levi Strauss, Chief Operating Officer Robert D. Haas handed out educational literature in the company lobby.

These actions were the outward signs of much more extensive initiatives. In several cases, the employee assistance program (EAP) played the pivotal role, providing both direct counseling and referral to community agencies. The company newsletters ran stories, periodically, to clarify benefits and policy. No, there was no risk from casual contact. No, one did not risk infection by donating blood. Wellness programs and health fairs included AIDS awareness and educational materials. Human relations or employee relations built educational support, including videos to be checked out and names of medical authorities who could speak to employee groups. The benefits function established working case management arrangements with carriers.

Without exception, each of these programs started from a moment of panic. An employee group had refused to work with a person rumored to have the virus. A person with AIDS returned to the workplace after a leave of absence. Employees were concerned about servicing telephones that might have been used by a person with AIDS.

EAST v. WEST—A LOOK AT THE CONTRASTS

New York City felt the impact of the AIDS epidemic earlier than San Francisco, and the number of cases in the New York metropolitan area continue to outpace those in San Francisco. The Coro Foundation, a not-for-profit research and public information agency, interviewed New York executives in the summer of 1987 and concluded that most were "avoiding the issue of AIDS." Asserts the Coro report: "Of all the institutional sectors [educational, religious, governmental] in the city, the corporate community has been the most silent."

Silence, however, does not equate with inactivity. Leon Warshaw, head of the New York Business Group on Health, served as adviser to dozens of major employers. His conclusion: "Most are doing a very effective job . . . in their own quiet way." Late in 1985, Warshaw's Business Group staged a day-long meeting for corporate leaders to share information and receive updates on legal matters. The New York Chamber of Commerce held similar events. More recently (and some 20 miles north of the city), Port Chester's United Hospital held a similar information-sharing meeting, drawing attendance from the major corporations headquartered in Westchester County.

Meanwhile, health educators from the New York City Department of Health and the Gay Men's Health Crisis (GMHC) were quietly conducting education programs for private and public sector employers. Most consisted of hour-long information and question-and-answer sessions, employee attendance optional. The New York Department of Health employs about 10 individuals for such training, with an additional complement of 15 dedicated to other forms of education (schools, street education in high-risk areas). The GMHC maintains a staff of more than 20 speakers, in addition to a corp of volunteers.

In virtually all cases, the corporate efforts are low-key and fragmented compared to the San Francisco approach. Yes, a few companies tell us that they have taken some action (New York Life, Citicorp, CBS, the J.P. Morgan Bank, Merrill Lynch, Time Inc., Bloomingdale's, and the Daily News, among others). Often, the action was as simple as a memo to department heads indicating that AIDS was covered under normal policy provisions for illness. Some firms earned an early reputation of understanding and support for persons with HIV infection (New York Transit Authority, AT&T, IBM, among others).

However, as the Coro report notes, there is no equivalent to the San Francisco Business Leadership Task Force, the 15 employers that pooled resources and created an ongoing mechanism for sharing experiences.

IS THE WEST COAST "MODEL" WORKABLE IN OTHER AREAS?

In San Francisco, the practice of reasonable accommodation came in the context of a municipal ordinance forbidding discrimination against persons with AIDS. The applicability of the Vocational Rehabilitation Act of 1973 plus various state statutes extending handicap definitions to AIDS now create a national parallel to that ordinance. Thus, "reasonable accommodation" in the best interest of the employer and employee, consistent with business objectives and the employee's ability to meet performance criteria, can be seen as a legal obligation.

Thus, some knowledgeable authorities suggest that the San Francisco lessons can apply to other areas. "Reasonable accommodation" means that an employee with AIDS will return to work after a period of convalescence. This heightens the need for education (companywide or workgroup specific), for clarifications of the company's position, and for other measures.

To determine if the San Francisco approach can serve as a "model," it may be wise to look, first, at the needs of those employee groups most likely to be affected, either directly or indirectly, by AIDS-related issues and problems. The effectiveness of any program might be judged by how well it answers concerns among such individuals. Such special groups might include the following:

1. Persons With HIV Infection Who Show No Signs of Illness

The New York City Department of Health estimates that about one-half million persons in that city now carry HIV, most of them between the ages of 18 and 49. Among persons in that age bracket, one in seven may carry the virus. To quote the Coro report, "It is no longer a question of if [employers] are affected or even when; they are now." On a national scale, the number of HIV carriers is estimated at between 1 and 1.5 million and is expected to climb to 2.5 million by 1991.

Sheila Lacy, AIDS training facilitator and caseworker for United Hospital in Port Chester, New York, tells the story of a patient who refused to allow his diagnosis to be communicated to his employer, preferring to cover treatment costs out of pocket rather than risk termination. Richard Snyder of Bank of America tells the story of an employee who called to say he was HIV positive and wished to resign, risking medical benefit loss rather than face ostracism—and this at a company with one of the earliest policies specifying coverage of AIDS and other "life-threatening illnesses."

The psychological needs of this group are intense. Daily life is marked with anxiety and fear—potential loss of friends, employment, and livelihood. The needs:

- Referral to both psychological and medical help. Early intervention can be important in prolonging life, according to doctors now treating such patients. The company's ability to facilitate such a referral at the earliest date can clearly make a difference in the person's workplace productivity and in the quality of care provided.
- Familiarity with benefits provisions: health insurance coverage, life insurance, leaves of absence, disability. Of course, the majority of these persons may not develop debilitating symptoms. There is, however, comfort in knowing what provisions are available.
- Other personal support: access to legal advice (wills), child care provisions, financial planning, and other family-related matters.
- Educational support for co-workers: if it becomes known that the person is carrying the virus, or should rumors to that effect begin to circulate.

Providing this support requires (a) that someone in the company is knowledgeable in such matters and has appropriate resources, (b) that the availability of such services is communicated within the company, and (c) that the company has established an atmosphere of trust and assurances of confidentiality.

The logical source is the director of the employee assistance program. Although EAPs were previously thought of as "large company only" services, smaller firms are reaping the same benefits through consortium EAPs and other shared-service arrangements.

HOW AN EMPLOYEE ASSISTANCE PROGRAM (EAP) CAN HELP

By Dr. Dale Masi

Originally started as occupational alcoholism programs, EAPs have broadened to become professional counseling services for employees with a variety of emotional as well as addiction problems. Many organizations now have EAPs, either in house or as a contracted service. Many are staffed with professionally trained mental health practitioners such as psychologists and social workers.

A company's EAP can become a calming voice amid reactions and over-reactions to AIDS-related incidents. The EAP can serve as a valuable resource in developing an AIDS policy, and it can help allay employee anxieties and fears as the company develops its philosophy and policy. An EAP is in a unique position to offer this kind of support, for several reasons.

Mental health professionals, especially social workers, have traditionally worked in hospitals where they have been the support for terminally ill persons, an experience that makes them particularly well suited to work with persons with AIDS and their families. Their professional objectivity enables them to operate as advocates for the person with AIDS while maintaining a perspective that allows them to remain sensitive to the anxieties of co-workers. EAP personnel will need to focus their efforts in two directions: providing appropriate support and referral services for persons with AIDS, plus educational information for co-workers who find themselves working closely with a person who has AIDS.

EAP professionals have been trained to work with community resources and are knowledgeable regarding those services. They will be able to assist in making contact with appropriate community programs. Employees with AIDS as well as their family members will need such support services to face the terminal illness, to make appropriate plans, and to achieve adequate medical care. Counseling from the EAP will give them the help they need to deal with other employees, supervisors, and the workplace in general. Family members as well as persons with AIDS will experience denial, upset, bewilderment, even shame, and they will need assistance as they face their own loss and try to cope with helping their loved ones.

The EAP can also help companies in the interpretation and implementation of legislation pertinent to persons with AIDS. The Rehabilitation Act of 1973, which declared persons with certain illnesses to be handicapped, has been interpreted as extending specifically to persons with AIDS. The EAP staff is already familiar with this law as it pertains to victims of other handicapping conditions and will be an invaluable aid in understanding its ramifications for employees who have AIDS, as issues concerning their disability benefits and termination on medical grounds arise.

According to Dr. Dale Masi, many EAPs provide 8 to 12 counseling sessions before making a referral to an outside counseling service. Masi recommends that, in the case of HIV infection, the EAP establish a more flexible and extended relationship. *All* counseling should be handled by the local office.

EAP directors now face the same moral dilemma as physicians: should the rule of confidentiality be breached if the well-being of others (a spouse, for example) is threatened? Some ethicists argue that the spouse *should* be informed. On the other hand, pragmatists would point out the legal implications of breaching confidentiality. This issue goes beyond the scope of this publication.

2. Persons Who Face the Problem at Home

Such persons may be experiencing denial, bewilderment, even shame. Some may not want the condition known to co-workers. They, too, need support and counseling on benefits, financial matters, and legal concerns. Again, an EAP can be the most valuable resource.

In addition, such people may need some workplace flexibility. Care for the family member can drain emotional energy and cause conflicts with schedules. The prevailing strategy of "keeping AIDS victims out of the hospital" may make these needs more prominent. The idea of case management is new to both benefits managers and employees. Clarifications of provisions may be needed. Note: The American Red Cross offers an 18-hour training session for family or friends who will be taking care of an AIDS patient at home, as well as an eight-hour orientation/preparation program for home attendants.

Suitable time off from work during the family member's last days is, of course, appropriate.

As with any personal tragedy, support from friends and co-workers plays a major role in helping the person work through bereavement. Such support systems develop naturally—unless curtailed by fear and misinformation. Again, the company's position can either hinder or help this natural support.

3. Employees Who Work with a Person Known to Be Infected

Some persons with HIV infection may want their conditions to be known. Should opportunistic infections become apparent, concealing the nature of the illness may become difficult.

Our 1985 publication told the story of a freelance word processing worker who had "gone public" with the information that he had AIDS, having appeared on a television talk show. His co-workers panicked. "They were lined up at the personnel office, asking what to do. One woman was hysterical—she was pregnant and was afraid I had left sweat on the keyboard." In this case, a co-worker took the educational initiative by contacting the Department of Health to arrange for a speaker and educational materials.

The story had a less than satisfactory ending. Although the educator completely calmed the hysteria, the company president ruled that the freelancer could work only after hours. "I guess I should feel lucky that I have any work," he told our researchers.

This story spurred responses from AMA members who had

similar experiences—a department, or, in several cases, an entire plant shut down for a short time because of similar panic.

The response: calls to both the legal department and to a "health educator"—someone with a medical background who was knowledgeable on the subject. In all cases, the educator proved more effective than the lawyer.

More recently, a human resources manager for a major food processing company told us that, to date, four employees had contracted HIV-related illnesses. "Contrary to our anticipations, the co-workers were caring and supportive." His advice: "Realize that first and foremost this is a human tragedy. The best approach is common sense—concern, sensitivity, support, caring. You don't need anything else." The company's "policy communication" consisted simply of memos posted on bulletin boards stating the "AIDS as any other illness" principle.

Most employers will find situations at either of these extremes or somewhere in between. The greater the potential disruption, the greater the need for policy clarification and access to readily available educational resources.

4. The Worried Well

This term is often used to describe persons who experience high anxiety because they fear they have been "exposed" to HIV. The fears are real, even if the presence of the virus isn't. Again, appropriate counseling through an EAP or referral to other resources provides the best answer.

Should the employer offer opportunity for employees to take the HIV-antibody test—with full assurance of confidentiality? About 6 percent of the 1,000 companies recently polled in an AMA survey offer HIV-antibody testing to their employees (See Appendix A). Most of these respondents represent service organizations wherein employees may come in contact with blood or blood products: health care organizations, fire departments, police departments, and the like.

A handful of other companies make testing available "on request." The practice would seem to make sense, from a purely theoretical standpoint. Voluntary testing would ease concerns among those who are needlessly worried and provide early intervention and referral for those who have contracted the virus. However, follow-up interviews with EAP directors and medical directors uncovered strong warnings against such practices: "The company should consider voluntary testing only if it is fully prepared to offer counseling both before and after the test." Nor, according to these respondents, should employers encourage the use of blood donation drives as a vehicle for testing, for the same reason.

5. Employees Accepting Assignment or Transfer to Certain Countries.

In some undeveloped or developing nations, health care practices are primitive by U.S. standards. Equipment used to extract blood or give injections may not be sterilized. The national department of health may not be screening the blood supply, or may have started such screening only recently.

The employer's liability when sending employees (and, in some cases, an employee's family) to such countries has yet to be addressed. Most assuredly, the person accepting transfer needs to be briefed on how to secure safe medical care. If the employee should return to the States for certain types of care, the criteria need to be worked out beforehand. None of these precautions guard against the need for unexpected emergency care. In some cases, it may be prudent for the individual to establish his or her own blood bank in the country.

Dr. Paul Goff, deputy medical director for the State Department, offers the following special considerations to employees with permanent or temporary overseas assignments:

- Do not have blood transfusions in foreign countries, unless the blood has been screened in American facilities.
- Do not take an HIV-antibody test in foreign facilities—confidentiality and anonymity cannot be assured.
- Avoid foreign medical facilities and other settings with a likelihood of exposure to disease and infection. This consideration is particularly relevant to members of risk groups and people with positive HIV-antibody status.

- Contact American authorities in individual countries to assess the local political climate. Newly emerging political ramifications related to the prevalence of AIDS in the United States may create certain hazards for risk group members and infected individuals. These situations change on a daily basis.

6. Employees Who Must Deal with Industrial Accidents

Yes, accidents do happen, and a supervisor may have to give some emergency care to a person who is bleeding.

The answer is the same for dealing with blood in health care settings, as specified by CDC guidelines. "Assume that all blood is infected." Use rubber gloves; clean the spill area with a mild solution of household bleach.

Such practices won't be adopted, however, unless the proper equipment is available in first-aid kits and supervisors are trained to use the equipment.

REACHING OUTSIDE THE WORKPLACE

Some company medical directors interviewed for this publication were quick to point out that the AIDS epidemic is truly an epidemic only for certain risk groups. As of this writing, the CDC demographic profiles show a significantly higher incidence of AIDS among blacks and Hispanics. Among Hispanics, the incidence is about double that of the general population. For blacks, the multiple is 2.3 times that of the general public. Many commentators have noted that poverty, drug abuse, and unemployment compound the problems of developing effective risk reduction and intervention programs for these groups.

As Caroll Curtis, director of medical affairs for Westinghouse, Inc., puts it, the company's position and educational program can have a "multiplier effect. . . .By teaching our employees, we influence their contacts. It helps get out the facts, rather than fear."

That influence may extend first to family members. (Some firms, including the J.P. Morgan Bank and IBM, allow employees to check out educational videos for home use.) Moreover, these employees are better prepared to talk with their children about risk behaviors, having first discussed the matter at the workplace. Second, the influence extends to social contacts as employees live their daily lives. Third, informed managers may wish to assume roles in community leadership. These individuals can bring both information and solid management skills to the community groups.

Reach outside the community may involve more formal ties. The Coro report urges New York City employers to "develop an AIDS task force through the New York City Chamber of Commerce and Industry."

> Once the task force is established, it would concern itself with educating the business community as well as applying its political clout to move AIDS education onto the public agenda. . . . Since business leaders play such a pivotal role in setting the city's agenda, a unified public statement . . . could provide a catalyst to promote the education of our larger community. Such a unified public statement would be acceptable to many companies that do not wish to make individual statements.

In addition, the Coro researchers recommend coordination of these efforts with public and nonprofit organizations. "By joining the consortium, the private sector would have immediate access to an entire network of expertise and information about AIDS education."

IS THERE A MODEL?

The appropriateness of any approach might be evaluated in the context of the foregoing discussion. Does the approach meet the needs of special employee groups, such as those defined? Does it help fulfill a larger social agenda?

The incidence of new HIV infection has begun to decline in San Francisco, the result of education that was both broad scale and targeted to specific risk groups. Although the corporate response cannot be given anything approaching full credit, it was clearly part of the solution.

The Levi Strauss, Bank of America, Wells Fargo approach—to the extent that the differences can be lumped together into a single category—may have succeeded because of a broad scale involvement: either an employee assistance program spearheading a drive that included education, policy formation, benefits revision, and the like, or an interdepartmental task force. The approach was comprehensive, addressing all the problems discussed above. Moreover, the company formed liaisons with community and civic groups—the San Francisco AIDS Foundation and the Shanti Project, for example. Money and educational materials flowed both ways.

Moreover, the firms were completely public about their response—everyone knew where they stood. Let's take a close look at two such employers.

BANK OF AMERICA

The decision to provide some type of information on AIDS arose in 1983, after an employee with AIDS returned to the workplace, according to Nancy L. Merritt, Bank of America's vice president, director of equal opportunity programs. At that time there was no formal policy with relation to AIDS, although guidelines had been distributed to personnel relations managers and human resources division directors. "In 1983, not very much was known about AIDS, and in fact, the environment was characterized by fear and hysteria," states Merritt. "Consequently, the company was reluctant to distribute a policy about the disease when there were still some uncertainties."

But by 1985, things had changed. Public awareness about how the virus is transmitted (and how transmission can be prevented) had increased, especially in the San Francisco Bay Area. It was at that time that the company sought to develop a strategy that included the creation of a written policy concerning AIDS, and other illnesses (that is, stroke, heart disease and cancer), under one policy related to life-threatening illnesses. A task force, consisting of representatives from personnel relations, legal, benefits planning and administration, and corporate health programs, was formed. After collecting information on AIDS, and life-threatening illnesses in general, the task force began on a five-point strategy to address the overall needs of their employees:

- Policy
- Benefit design
- Communication
- Training
- Support and referral

1. ***Policy***. The policy statement: "Employees with life threatening illnesses may continue to work as long as they meet acceptable performance standards, and medical evidence indicates their conditions do not pose a health or safety threat to themselves and others." If warranted, the company will make reasonable accommodation for employees with life-threatening illnesses consistent with the business needs of the unit. The policy also includes a list of services provided to employees with life-threatening illnesses and guidelines for managers to use in handling specific situations. Various department heads and members of the senior management group were given opportunity to review and comment on the document before it was put into final form. (Note: The full policy is reproduced in Chapter 3.)

2. ***Benefit design***. Incorporating medical case management—allowing for medical care in the home or a hospice and providing chore services—became a key component for containing health care costs. This is consistent with the San Francisco community model for treating AIDS patients, where most treatment is received either through home health care support or through the assistance of a hospice. As a consequence, per-person health care costs were running $30,000 to $40,000 below those in other cities.

3. ***Communication strategy***. The policy, when developed, was distributed to each manager, who in turn reviewed the policy and procedures in staff meetings. BankAmerica also made the policy available to other interested employers and community representatives. In addition, various articles about the policy, as well as those addressing the virus, were published in the employee newsletter, *On*

Your Behalf. Articles appearing in the January 1986 and July 1987 issues of that publication provide basic information about how the virus is spread, answer commonly asked questions about benefits, and reaffirm the points articulated in the policy. For example, the July article, "Living with AIDS," clarifies the protection offered through accrued benefits in the company's long-term disability plan. Coverage for health plans (medical, dental, vision) continues at the predisability rate as long as the employee continues to make monthly contributions. Basic (employer-paid) life insurance continues as does employee-paid life insurance benefits as long as the employee makes monthly contributions.

The same story reports on the response from employees with AIDS—that the support they have received "played a key role" in their return to comparative good health after an initial convalescence. "They all feel that the opportunity to return to work and be productive was important to them both physically and psychologically."

Stories such as these have become an important part of the communication strategy. In addition, a number of videos and brochures are available to those employees and managers who want additional information. Anyone seeking additional information is invited to contact personnel relations or corporate health programs.

*4. **Training***. AIDS training and education at BankAmerica is kept consistent with other programs available for a variety of human resources and health issues. To do this, the company provides general information through the employee newsletter and staff meetings, and more specific education in situations such as an employee with AIDS rejoining the workforce. In addition, prior to the announcement of its policy, BankAmerica's personnel relations specialists received special training that addressed the disease, the policy, medical coverage, and communication techniques. This training is now conducted, at least yearly, for all personnel relations specialists, to keep their knowledge current.

*5. **Support and referral***. BankAmerica has developed a support and referral capability to direct employees with life-threatening illnesses to community self-help and support programs, for psychological and social support. In addition, their corporate health programs department provides a directory of national telephone numbers for information on all major life-threatening diseases. This is available, on a confidential basis, to any employee.

Going One Step Beyond

In March 1986, BankAmerica, as part of its membership in the San Francisco Business Leadership Task Force, helped develop a program for health education at the workplace consisting of films and other materials. The task force, made up of 15 major employers in the San Francisco area, made these materials available to all Bay area companies that attended an AIDS conference, organized by the group. After the conference, all materials were given to the San Francisco AIDS Foundation for distribution. All revenue gained from the sale of these resources helps the foundation provide future programs.

Lessons from Bank of America

In November 1987, AMA researchers followed up by asking Ms. Merritt to put her work into perspective. A veteran of AIDS education, Merritt stressed the importance of a policy in terms of both employer and employee protection—with trust as the bottom line. "The importance of having some type of policy, or guidelines, cannot be overstressed," asserts Merritt. "If you don't have a policy in place, you are looking very much at the downside. . ." that being fear (which can cause workplace disruption), undetected illness and anxiety, uncontrolled health care costs, and discrimination lawsuits.

To set up that policy, Merritt suggests that you look at the resources you already have in place—in your human resources department, compensation and benefits area, communications, and legal department. Then formulate a policy and program that best suits your organization.

But trust is the bottom line. "AIDS is no longer the other person's problem. And it is no longer an obvious problem. Persons with HIV infection who show no signs of illness need to be addressed in the corporate setting. In order to help them, you have to set up an environment of trust, where they can come to you sooner and sooner in the process of the disease. Education and a visible, well understood policy are important elements in developing that environment of trust. It is a difficult job to get trust without it."

PACIFIC BELL

In 1985, Dr. Michael Eriksen, director of preventive medicine for Pacific Bell, made this statement to an AMA researcher: "We don't believe in having an AIDS policy. We want to treat AIDS like any other medical disability. There's no doubt that it's a very serious health problem, but I don't think it requires any special handling. Insurance-wise, it falls under our current benefits plan." And according to Dr. Jackie Wood, Pacific Bell's current medical director, the same response still applies. Adds Wood, "Since 1985 there have been no serious incidents involving AIDS patients and other workers or the union."

Statements such as these, plus the company's philanthropic contributions to AIDS research and education, plus a series of widely cited stories in the company newsletter, leave no doubt about the company's position. That position has become a de facto policy, and company spokespersons will occasionally lapse into talking about "our policy" even though there is no official document in the handbook.

Pacific Bell has focused its efforts to address three concerns:

- Employees with AIDS who are concerned about their lives and their jobs.
- Employees without AIDS who are concerned about contracting AIDS from employees, customers, or others with AIDS.
- Employees who are concerned about friends or relatives who have or are dying from AIDS.

In a speech given in September 1987, Terry Mulready, vice president, corporate communications, indicated that the "primary challenge presented by AIDS" is in the hands of managers. Three principles should govern their responses:

1. Education is the key to effective management.
2. Managers must maintain a discrimination-free workplace.
3. In dealing with AIDS, managers must take all reasonable steps to keep their workers as healthy as possible.

Pacific Bell's education encompasses a wide range of services: speakers, seminars, videos, brochures, availability of in-house medical counselors, a manager's guide, and information periodically published in their company newsletter, *Update*.

USING THE COMPANY NEWSPAPER

The material in this box is excerpted from *Update,* the Pacific Bell in-house newspaper. The first story, drawing its scenarios from concerns expressed by employees, appeared in 1985. Subsequent articles provided factual information and some medical advice. The John Zorbas story appeared in May 1987. Used here by special permission from Pacific Bell.

AIDS

In a Contra Costa office, Jane sits at her desk processing service orders. Jim, her co-worker of several months, slides a form onto her desk. Glancing up, Jane is startled to see purple blotches on his arm.

The mental note is etched.

As far as she knows, Jim is single, though he rarely talks about his personal life. Jane knows he just bought a home with "a friend" and remembers that he'd seemed relieved to get out of the city and settle down.

On the train home, Jane grows anxious. She thinks of her family. What if Jim has something contagious? Could she catch it and what if it can't be cured? Why should she be subjected to risks?

In Los Angeles, George had never had any problems with his coin collecting job. But recently, one of the bars along his route had closed. The owner died rather suddenly and George was shocked. He'd seen the guy several times—though not in the past few months. He'd seemed fairly young.

After the bar closed, George decided to call his supervisor. He'd been getting nervous about going into that bar and other ones along his route, about handling the phones. Once, George saw someone passing out AIDS pamphlets. Though curious, he wasn't sure he should touch even those.

Richard spent a week in Hawaii trying to shake the fatigue, fever, and swollen glands that had plagued him for the last few months. When he returned to his Stockton office, tan and thin, everyone thought he looked better.

But, rather than feeling rested, Richard could barely drag himself to work. After arriving at the office late several mornings in a row, Richard faced the fact that he had to see a doctor.

What he heard sent chills down his spine and drained the color from his face. The diagnosis was AIDS.

Offering what consolation he could, the doctor gave Richard the name of the nearest AIDS counseling center, and suggested he move closer to San Francisco, where treatment might be more advanced.

His immediate symptoms could be treated and, for the time being, he'd be well enough to return to work. Richard wanted to work. He didn't want to stay home, alone with his illness—to admit to himself he was very sick.

But, returning to the office was overwhelming.

As a gay man, Richard had known it would be difficult when he took the promotion and transferred, alone, to a smaller, unfamiliar town.

People at the office had been surprisingly friendly, yet, while casually socializing with his coworkers, Richard had never let it be known that he was gay.

Panic gave way to depression.

The scenarios above are not the experience of three particular individuals but rather a compilation of the ways AIDS can affect employees at Pacific Bell. They're a starting point for the questions that arise from people everywhere, across the state and around the country.

With 75,000 employees from all walks of life, Pacific Bell is a city of sorts, sharing many of a city's concerns for the financial, psychological, and physical well-being of its inhabitants. With the AIDS epidemic gradually affecting more and more people in a variety of ways, only knowledge and understanding can separate myth from fact, reduce the undercurrent of anxiety, and lessen the problems encountered by the individuals with the disease and those around them.

What is AIDS?

AIDS stands for Acquired Immune Deficiency Syndrome. It is a disease caused by a virus that attacks the body's immune system, leaving it vulnerable to certain types of infection.

But it isn't the disease called AIDS that kills directly.

John Lorenzini, founder of People With AIDS Alliance, explains, "The infections that develop are what kills, not the AIDS virus itself. So, what the person with AIDS exhibits are the symptoms from these infections—not one specific to the disease called AIDS."

Due to the number of infections, the AIDS symptoms vary—fever, night sweats, swollen glands, unexplained weight loss, yeast infections, diarrhea, persistent coughs, fatigue and loss of appetite. (It should be noted, of course, that these symptoms can indicate many different illnesses—not just AIDS.)

Two infections have appeared more frequently than any others in people with AIDS: *Pneumocystis carinii* pneumonia, a lung infection caused by a parasite, and Kaposi's sarcoma, a rare form of cancer.

They are not usually found in people whose immune systems are normal.

Who is Likely to Get AIDS?

As a relatively new disease, AIDS statistics are constantly changing. Research turns up new evidence all the time, but findings show consistently that the number of people stricken has been multiplying. Many sources agree that this may only be the beginning.

As of April 1985, physicians and health departments in the United States reported 10,000 patients meeting the "surveillance" definition for AIDS. Although the first case was reported in the spring of 1981, over half have been reported within the last 12 months.

According to the U.S. Department of Health, AIDS cases are divided into the following groups:

- Homosexuals/bisexuals (73.4 percent of the total number stricken)
- IV drug users (17 percent)
- Transfusion recipients (1.4 percent)
- Heterosexual contact (0.8 percent)
- Hemophiliacs (0.7 percent)
- Others/unknown (6.7 percent)

Current statistics show that 90 percent of the stricken adults are between the ages of 20 and 49; 94 percent are men.

Commenting on these figures, Jackson Peyton, Education Director of the San Francisco AIDS Foundation, points out, "It's important to understand that AIDS is not just a gay man's disease. It's affecting a broad spectrum of people.

"Historically, it has not been confined to the homosexual population. We know that in Africa, AIDS is spread through heterosexual sexual contact. So AIDS is everyone's concern."

"ONE EMPLOYEE'S PERSONAL STORY"

John Zorbas, age 34, was diagnosed as having AIDS at the end of January. Originally from Minnesota, he joined the company eight years ago after earning an MBA at the University of San Francisco. John started in computer operations and went on to work as a systems analyst. For the past two years, he has been working as a second-level manager in Internal Auditing. John had been diagnosed with ARC (AIDS-Related Complex) in October 1985. Also a diabetic, he soon began to have health problems, including failing eyesight. He first informed his supervisor of his condition in mid-1986.

On April 1, 1987, he left the company on disability.

"I used to be involved in a lot of things—I was president of the Golden Gate Business Association, very active on KQED's Community Action Panel—but I had to start cutting back on those activities. My energy level was going down and I didn't want my job performance to slip.

"I told my boss, Patty, months ago and she's been very supportive. She's the type of person you can talk to about anything. It was very important that we could be really open with each other. It made things a lot easier.

"A lot of it has to do with your boss. I hope all managers make sure they're educated about the facts. I think there's still a lot of employees who come in from the suburbs and aren't that educated. They still think AIDS is just some gay disease.

"I chose not to tell my co-workers, except those I'm close to and would have told anyway. If I had cancer, I wouldn't have broadcast it either. I see it as a very personal issue, and I didn't want them to treat me any differently than they had been. I wasn't afraid they would treat me differently because they were afraid of catching AIDS, but because they were concerned about me.

"Sure, I need help with certain things now, but no big deal needs to be made out of it. I don't want special treatment, people hovering over me always talking about it. I want to be treated as normally as possible.

"Work has always been important to me, so I hated to leave. But most of my job is reading and writing and my eyesight is pretty far gone. I don't sit around a lot at home though because I'd go nuts. I try to go out to lunch, get out and do things. But I miss being involved at work, seeing what's in my inbox, keeping up with what's going on.

"The company's made things easier by providing resources, helping me get all the benefits forms filled out, and I haven't had to worry about losing my benefits or health coverage. It's stressful enough to have the disease; you shouldn't have to go through all the other stresses that can go along with it."

How do you sell an AIDS program to top management—and to your employees? According to Mulready, Pacific Bell has identified five steps to help influence the company to do "the right thing."

> First, realize that in your push to establish an AIDS program, you have allies throughout your company—some of them officers and directors. You need to identify these people and work with them.
>
> Second, you need to stress that having an AIDS policy is the right thing to do. Our employees, our companies, our communities, our country face a problem—and there is no excuse for inaction.
>
> Third, having said that, you need to appeal to the self-interest of relevant departments within your company.
>
> You need to convince your benefits group that AIDS education holds down health care costs. And you need to convince Contributions that fighting AIDS is a good social investment.
>
> Fourth, you must make people aware of the fact that AIDS is an unusual opportunity to do well by doing good. The resources you devote to AIDS will more than pay for themselves in savings on future health plan expenditures and public and employee recognition. People want guidance—and if you give it, you'll be the leader.
>
> Fifth and last, people love favorable recognition. Few issues get more [recognition] than AIDS, and if you're on the right side of an issue, favorable attention will follow. Such attention must never be the reason for starting a program, but it can be a tool to motivate otherwise hostile or indifferent people.

Much of the company's publicity has come as the result of community involvement. Since 1983, Pacific Bell has contributed to many organizations, universities, and programs that have helped promote

AIDS awareness throughout California. Some of the most notable:

1985-1986: $7,000 to the University of California Medical Center in San Francisco to cover the cost of printing two editions of *A Resource Manual for Persons with AIDS.*

1986: Pacific Telesis Group participated in the Business Leadership Task Force conference on AIDS. Other corporate participants included Bank of America, Chevron, Levi Strauss & Co., Mervyn's, and AT&T. The conference generated a comprehensive program to educate people about AIDS in the workplace. For its part, the Pacific Telesis Foundation gave $35,000 to the San Francisco AIDS Foundation to produce a videotape called *An Epidemic of Fear: AIDS in the Workplace.* The tape is serving as an educational and fund raising tool for the foundation.

1986: Pacific Telesis Group contributed $5,000 to the "No on 64" campaign. Ballot Proposition 64 sought compulsory testing for the virus and the quarantine of people with positive results.

1986: Pacific Telesis Group helped sponsor a one-day conference on grant makers and AIDS; the Pacific Telesis Foundation donated $6,000 to produce a videotape about the conference called *AIDS: A Community Response.* The video, produced by Pacific Bell's Corporate TV group, helps grantmakers better understand the issues raised by AIDS.

November 1987: Pacific Bell and the Pacific Telesis Foundation launched a first AIDS education program for Hispanics. Pacific Bell underwrote the cost of air time and promotional spots for a two-hour statewide TV special that appeared on the Spanish language Univision network. The Foundation donated $7,500 to produce a novella (or drama) which formed the first hour of the special, and 5,000 copies of a companion book on AIDS. The foundation also gave the American Red Cross $20,000 for a hotline and an education program launched during the television special.

2

Dealing with Workplace Problems: Questions and Answers

This chapter presents consensus answers to a series of questions asked by members of the AMA research staff. Respondents included providers of educational services, managers who had dealt with workplace problems, company medical directors, and AIDS patients who were working at the time of their diagnosis. Respondents were selected on the basis of first-hand experience with AIDS-related situations. Consensus answers as written in 1985 have been refined on the basis of reader responses and similar question-and-answer research conducted by other organizations to whom our research staff served as consultants. A number of questions and their respective answers have been added, again on the basis of reader responses.

What Should You Do If a Person in Your Workplace Does Have AIDS?

The issue is a highly sensitive one. As Jackson Peyton, educational director of the San Francisco AIDS Foundation explains, "A person may have a lifestyle that co-workers don't know about. So there is the issue of being gay or a drug user, compounded with the fact that people don't know how to deal with someone who has a terminal illness.'' As a manager you want to protect the privacy of the individual. If the individual requests confidentiality, that request must be honored.

As a number of respondents noted, "The manager of a person with AIDS sets an example for the behavior of others.'' The following actions were suggested.

1. Offer the person support. Tell him or her that there are still a lot of misconceptions about the disease and that you will talk with anyone who voices concern. At this point, clearing the air and dealing with the facts is the best course for everyone.
2. Begin immediately to educate yourself if you haven't done so already. Know the facts so that you can answer your employees' questions or find out the answers for them.
3. If your organization does not have an AIDS education program in place, use what influence you can to initiate one. An article in your company newsletter, such as the one published in Pacific Bell's *Update,* could be the beginning of a company-wide educational effort. Levi Strauss set up an information booth in its lobby and COO and President Robert D. Haas handed out literature personally. Note: Many companies have waited until they have had an incident of AIDS or a rumor thereof before gathering resources for such a campaign. Our respondents spoke with a single word on one point: "Don't wait until you have your first AIDS patient.'' As organizations gain perspective, they are becoming proactive.
4. Prepare to provide special education for the workgroup by making a request to the appropriate department, by obtaining literature to hand out, or by obtaining the name of a local medical authority who can talk directly to your workers. Your local chapter of the American Red Cross, GMHC, hospitals in your area, and the state or city department of health are often the best sources for knowledgeable speakers.

Isn't Confidentiality Difficult, Perhaps Impossible, in Some Situations?

Note: This question arises because of the progressive and episodic nature of HIV-related illness. At some point in time, it may be impossible to conceal the nature of the illness; decisions regarding workplace accommodations and, potentially, disability will have to be made. It is likely that more than one person will be involved in the decision making. At the same time, all employees have a "right to privacy," as guaranteed by the Fourth Amendment to the U.S. Constitution and by ordinances in various states.

Our respondents emphasized two points.

- Where there are no signs of illness, there is no reason, need, or justification for disclosure. Most of the conditions associated with ARC happen to people with perfectly healthy immune systems; again, there is no need to disclose information.
- In dealing with situations where there are signs of AIDS, respondents emphasized the human element: "Just use a common sense approach—be sensitive, concerned, supportive." "Be aware of the employee's needs and the needs of the workgroup." "Handle it as you would any other serious medical condition."

Should Education Be Targeted to a Specific Group When, Say, a Person with AIDS Returns from a Leave of Absence?

Some respondents urged caution if the company did not have an AIDS education program in place. Be sure the person with AIDS gives his or her consent and has considered the fact that attention may be drawn to his or her workgroup.

A number of West Coast firms made education for the workgroup standard practice when an individual with AIDS returns after a leave of absence. Reports of this in prior AMA publications raised concern about confidentiality; legal consultants argued that if such practices were adopted, the company would be well advised to obtain a signed release from the individual.

We put these questions back to the companies that practice group specific education. No, they did not, as a matter of course, obtain a written release. They did, however, talk frankly and supportively with the individual. It invariably happened that the person wanted his condition known, welcoming opportunity to clear the air.

The practice speaks well for the power of education. Still, there are conditions in which a signed release may be in order. The company culture may dictate the tack to take. Less public, one-on-one discussion may sometimes be more appropriate.

What Happens If the Workgroup Refuses to Work with the Person with AIDS?

One can cite dozens of cases in which this has happened. In most situations, the employer used both a legal consultant and a medical education consultant. In the end, the educator proved most effective. In none of the case histories reported by our respondents did this tactic fail to take the fuse out of a potentially volatile situation. In two cases, however, concerns lingered among the management group long after the employee's co-workers overcame their aversions.

In some instances, co-workers took responsibility for providing the education; in one case, a person with AIDS obtained a hundred pamphlets and asked that his vice president distribute those to people in the company.

Although education on how AIDS is transmitted has been broadly aired in the media, many people still feel that casual contact with a person with AIDS poses a threat. As one respondent pointed out, people may fall into two categories:

- Those who don't know the facts and will be reassured once properly educated.
- Those who have "heard" the facts, but still don't believe them, because they mistrust authority (doctors, government, or management) and believe the "real facts" are being withheld.

This second group is hard to reach. Fears are often mixed with homophobia and are based on deep-seated prejudices, respondents noted.

The managers among the respondents expressed the need to protect the interests of all their employees. People who resist the facts may be acting in "good faith"—they may truly believe that casual contact poses a risk.

What if the education effort fails—that is, a employee refuses to work with a person with AIDS? Our respondents indicated that they would have no choice but to apply disciplinary action within the framework of company policies for attendance and job performance.

If the Work Setting Presents a Risk for Minor Cuts, Shouldn't at Least the Supervisor Know if a Person is HIV Positive?

It is now standard practice for health care providers to treat all blood as if it were from a person with HIV infection. About 1.5 million Americans may be carrying the virus—one person in 30. Most show no signs of illness.

Of the tens of thousands of health care providers who have cared for AIDS patients, about a dozen have contracted the virus in the process. In all cases, there was a lapse in clinical technique—a failure to use gloves. The chance of infection is, indeed, remote. But the risk (however small) is there.

In its "guidelines for other workers sharing the same work environment," the CDC has stated, "Equipment contaminated with blood or other body fluids of any worker, regardless of [HIV] infection status, should be cleaned with soap and water or a detergent. A disinfectant solution or a fresh solution of sodium hypochlorite (household bleach) should be used to wipe the area after cleaning."

What Kinds of Accommodations Should Be Made?

The law requires "reasonable accommodation" for a person with AIDS. The following principles are worth considering, according to our respondents.

1. The person must be able to perform the essential tasks of a job.
2. "Reasonable" depends on the resources available to the employer, including such factors as the size of the workforce, the

nature of the industry, the requirements of the job itself, and the cost of making accommodations.

3. Be aware that the person with AIDS has a damaged immune system and may be susceptible to viruses and health hazards in particular work settings.
4. Most often, companies offer modification in hours, such as flextime or part-time work at a reduced salary, or allowing the person to begin the day at a later hour; job changes such as reduced workload for a lower salary, job sharing, or a shift to less physically demanding work; and installing special equipment or eliminating architectural barriers to accommodate a wheel chair, for example.
5. If the individual has a poor performance record, objective measures of performance and carefully written documentation are imperative. Should termination be necessary on the basis of ability to perform the job (separate from the issues of illness) the company will want to be able to support its decision.
6. Under the law, the employer may consult with the person's physician to devise appropriate accommodations.

Two respondents reported on morale problems. Employees felt their workload had become unfair because the person with AIDS had a diminished capacity to contribute. Moreover, working with any person who shows signs of illness can dampen morale. In both cases, the respondents' attempts to air the problems and solicit everyone's help created a spirit of camaraderie.

The disease of AIDS has a variable course. A person with AIDS may feel well for extended periods of time. Often he or she chooses to continue working, not just to assure income and benefits but because it enables him or her to maintain the semblance of a normal lifestyle. In our original research we found that a number of the AIDS patients interviewed simply requested, and received, disability as soon as the diagnosis was known—and in some cases, before the condition was diagnosed. The stories are compelling. Consider this from a middle manager at a New York City tour company: "Those who have worked all their adult lives and have great pride in their accomplishments . . . suddenly find themselves depending on government disability checks, food stamps, and Medicaid."

More often employers do come to an arrangement whereby the person with AIDS stops work but continues to receive medical benefits. This is often the most expedient way out. But BankAmerica urges in its guidelines: "Be sensitive to the fact that continued employment for the employee with a life-threatening illness may sometimes be therapeutically important in the remission or recovery process, or may help to prolong that employee's life."

What if the Employee Can No Longer Do the Job?

Consensus among respondents and guidelines from other commentators suggest the following approach: If you notice a decline in performance, first ask if there is anything you can do to make the job easier. As the disease runs its course, it is inevitable that at some point the employee will no longer be able to work. Reasonable accommodations can prolong the employee's productive period. You can ask how the employee feels about it and, together with the personnel department, seek medical consultation about whether the employee is fit to work.

Keep written documentation of performance.

When separation becomes necessary, find out about the employee's eligibility for medical benefits and disability, and explore the other alternatives available. The University of California's termination policy states:

> Separation may be pursued only if an employee can no longer perform the essential functions of the job even with reasonable accommodation, if the employee has exhausted sick leave and any personal leave, and if the separation will not jeopardize the right to apply for disability benefits provided by the employee's retirement system.

Can We Refuse to Hire a Person with AIDS?

Our respondents were clearly aware of the legal framework: If a candidate is otherwise qualified for the job and is physically able to do the work, refusal to hire on the basis of AIDS would constitute discrimination against a handicapped person. Jobs that require an individual to be AIDS-free as a condition of employment are few and far between.

As a group, homosexuals are not protected under Title VII or under most state fair employment laws, although many large cities have local laws forbidding discrimination. Thus, although discrimination on the basis of sexual orientation may not in itself be illegal in many jurisdictions, refusal to hire a homosexual man, for example, because of the possibility that he may now or at some time in the future contract AIDS falls under the handicapped or "perceived" handicapped provision of the 1973 Federal Rehabilitation Act as well as under most state laws. "Perceived" handicap has the same protection as actual handicap.

The University of California Personnel Policies and Procedures with Regard to AIDS spells out its nondiscrimination policy:

> At present, it appears that AIDS/ARC is substantially more prevalent among male homosexuals than any other population subgroup. As a result, there may be a tendency for fear of AIDS/ARC to express itself as homophobia. University policy prohibits discrimination in employment based on sexual orientation and provides for resolution of discrimination complaints through the applicable grievance or appeal procedure.

How Do You Handle Rumors that Someone has AIDS or ARC?

Respondents from companies without education and support initiatives in place had difficulty answering this question. Several indicated that a company medical director had provided valuable help.

Firms with education programs and support liaisons in place indicated that a more direct approach could be used: Tell the individual about the rumors, point out that there are still misconceptions, offer support, and so on.

If the person does not have AIDS or does not want to discuss it at work, one still has the rumors to deal with. Respondents suggested handling such situations with one-on-one discussions with the people involved.

Several respondents emphasized that hearsay and rumors are not grounds for impetuous action. ARC is difficult to diagnose because the symptoms parallel those of other illnesses.

Can an Employer Ask a Job Candidate to Take an AIDS-Antibody Test?

Respondents often answered with another question. "What would the company do with the results, once obtained?" Withdrawal of an employment offer on that basis would clearly be discriminatory. As the Business Leadership Task Force pointed out, "It should be kept in mind that once an employee has requested an HIV-antibody test, it could be difficult to prove that an individual was *not* discriminated against because of the results of the test."

3

The Role of Corporate Policy

One of the largest U.S. food processors has put policy adoption and companywide training on hold, in part because of the "irrationalities of the marketplace." (A policy of "reasonable accommodation" and press pick-up of the training might signal that the company "had a problem." How might this influence customers?) The company sends representatives to AIDS-related business forums, but the representative wears a badge carrying a fictitious company name.

To Surgeon General C. Everett Koop, widespread adoption of corporate policy would be a key weapon for fighting what he terms the "epidemic of ignorance and irrationality." American business is known worldwide for its pragmatism. Thus, many feel, a unified statement from corporate America—facts rather than fear, direct confrontation of the issues rather than retreat—could become a cornerstone for other effective action. Never before in the history of modern epidemiology have fear and ignorance so curtailed the processes of detection and diagnosis, thus blocking early intervention and treatment, counseling, and education aimed at behavioral change. Nor within the current emotional context can anyone hope to measure the extent of infection via nationwide testing. Anyone who believes that such test results could be kept confidential is either "naive or outright deceitful," as Cardinal O'Connor has pointed out.

The reasoned corporate response vis-a-vis public sentiment surfaced in response to our 1985 edition of *AIDS: The Workplace Issues*. That edition contained a case study of a prominent New York City restaurant where four waiters had died of AIDS-related illnesses. Our corporate readers accepted the case study for its intended purpose, and a dozen companies in food-related businesses called to ask about specifics and share similar anecdotes. When the *New York Post* picked up the story as part of a series on how city employers were responding, we received a much different call: people were outraged that an organization such as the American Management Association would create a public menace by not disclosing the name of the restaurant to the Department of Health.

Specifically, the restaurant was dealing with the problem first, by ensuring that normal procedures were followed. "The restaurant has a very strict policy when it comes to cleanliness," the general manager told us. "Before waiters can leave the bathroom, they must always wash their hands. And none of them are allowed to touch food with their hands. This is our normal policy—it's not one we've implemented because of AIDS.

"There is no need to take special precautions in service. But I'm not sure our patrons know that—or believe it. And so we have to be careful; we have to keep quiet about it. But if someone decided to come out with it, that they had AIDS, I think we'd have to force them to leave. Letting them stay on the job would probably ruin us."

Thus, for the general manager, the granting of full medical leave was the only viable solution. Might the company simply work through a carefully concealed discrimination—refuse to hire persons who might be at risk? "That kind of discrimination just wouldn't be right," said the general manager. "Second, most of our employees are gay and about 80 percent of our applicants are gay. We'd lose our workforce." Without doubt, there are situations in which the adoption of a formal, AIDS-specific policy of reasonable accommodation is problematic. More on this later.

WHAT SHOULD AN AIDS POLICY INCLUDE?

Surgeon General C. Everett Koop

I cannot be exhaustive in my suggestions of what should be included in your company's AIDS policy, but let me suggest at least the following:

- Treatment of AIDS should be within existing policy for illness.
- Employees with AIDS should be offered the opportunity to work as long as they can—bearing in mind the risk to the person with AIDS in some high-risk areas. . . .
- AIDS patients should perform satisfactorily or be offered lesser responsibility.
- Employees should not be granted transfer requests inconsistent with other transfer policy because there are persons with AIDS at their worksite.
- Confidentiality of health records of persons who carry the virus but are not ill must be maintained.
- Respect for the individual—consistent with the company's experience—must be stressed by management and subordinates.
- An educational program must be implemented.

Excerpted from an address given at the Allstate Foundation Forum on Public Issues, Chicago, Illinois, October 13, 1987.

EXHORTATIONS TO ADOPT POLICY GO UNHEEDED

Dr. Dale Masi, the nationally recognized authority on planning and implementation of employee assistance programs (EAPs), emphasizes that policy questions—especially as they relate to hiring, testing, health benefits, and changes in job category—will become more pressing in the years ahead. According to Dr. Masi, "AIDS is a problem that must be dealt with; the head-in-the-sand approach will not work. . . The naiveté of some corporations is exemplified by a major automobile manufacturer who said that there was no reason to believe that any of their 382,000 employees worldwide had AIDS. I know of an airline that felt certain none of their employees had AIDS. Yet according to their employee assistance program director, there were at least 14 cases in that company."

To Dick Haayen, chairman of the board of Allstate Insurance Company, corporate social responsibility in dealing with the AIDS epidemic is now inescapable. "Business as usual includes a commitment to corporate citizenship of the highest order," Haayen told a group of more than 100 business leaders at an Allstate-sponsored symposium in October 1987. "We have challenged ourselves to help bring about constructive change in the communities where we live and work. And we have dedicated ourselves to being as effective on the social side as we are in our business operations." The Allstate Foundation initiative, under the banner of "AIDS—Corporate America Responds," has produced the recently published white paper on a variety of employment issues, including benefits, policy formation, and education.

Why involvement from corporate leaders? First, according to Haayen, action from the corporation spills over into the community via employee involvement. Second, education programs at the worksite reach families and acquaintances. Moreover, "companies do represent a tremendous pool of talent. They do possess significant resources. And often, they bring to the task a results-oriented, can-do approach—attributes that are just as effective in pursuit of social objectives as they are in business."

A poll of 600 senior executives conducted by Louis Harris and Associates, Inc. (in November 1987) found that only about 11 percent of the respondents had specific policies for dealing with AIDS, about double the number counted by a similar American Management Association/*Personnel* magazine survey in 1985. Only 15 percent of the respondents to the Harris poll indicated that their companies had education programs—only a few percentage points higher than the 1985 AMA/*Personnel* survey. Moreover, 72 percent of the Harris poll respondents without education programs indicated that they had no plans to implement them.

WHY NO POLICY?

First, one must point out that the absence of policy does not equate with the "ostrich syndrome." Leon Warshaw of the New York Business Group on Health often states, "Many companies are dealing very effectively with AIDS in their own quiet way"—and this from the perspective of having worked as consultant and educator to dozens of major employers in the New York metropolitan area. On the West Coast, Levi Strauss—often cited for its exemplary approach of education and referral—operates without a formal written AIDS-specific policy.

AMA interviews with companies not having AIDS-specific policies shed some light on the inertia. The factors mentioned, in descending order of frequency:

1. We've yet to have a case of AIDS; we'll deal with the problem if and when it surfaces.
2. Since the prevailing wisdom is to treat AIDS-related illness as one would any other chronic or life-threatening illness, no specific policy is needed. Why a policy on AIDS, but not cancer, or cardiovascular illness?
3. The few cases we've had were dealt with on an individual basis—the situations were too sensitive to do otherwise.
4. We understand that the virus can't be transmitted casually, but our customers don't.
5. An AIDS-specific policy opens the possibility that the company would become a haven for HIV carriers seeking medical coverage.
6. Medical information changes daily. Thus, any policy would have to be temporary.

The last concept (changing medical information) is the easiest to counter. Indeed, some of the information does change. The "conversion rate" (the percentage of HIV carriers who will develop opportunistic diseases, and in what time frame) is the subject of ongoing revision. The distribution of diagnosed cases by standard demographics and likely cause of infection changes with each new CDC report. Accurate measurements of how many persons carry the virus have yet to be made. However, medical information about how the virus is transmitted has not changed substantially since 1985; nor is it likely to change. (The number of cases classified as "unknown transmission" has decreased from 5 percent to 3 percent. These are cases in which the person with AIDS has refused to answer questions about personal sexual activity or drug use.)

None of the changes cited above impact directly on company policy. An employer's response to reasonable accommodation is a matter of both legality and humane concern, not case distribution and conversion rate.

Also easy to counter is the idea that policy is not needed because the company has not yet had its first case. That "first case" may uncover more panic and anxiety than good business practice can allow. More important, those who articulate this stand may be overlooking the needs of special employee groups: those who carry the virus but are not ill, the employee who must care for a person with AIDS at home, employees accepting foreign assignment, and the like. *Quite likely, the company already has employees in one of these categories.*

Might a company become a haven for AIDS victims? We put this concern, as expressed by some of our respondents, back to those companies that do have AIDS-specific policies. Most (to put it frankly) were perplexed. Well . . . no . . . that isn't a concern. We followed with additional questions:

> Do you feel your company is in a better position to deal with AIDS-related issues because it has a policy? If yes, how so?

The answer to the first question: an unequivocal yes from all respondents. Why are they in a better position? "Problem avoidance"

Connecticut School Board

In 1986 the school district of Greenwich, Connecticut, implemented an AIDS policy for teachers and students. The first of its kind within the U.S. public school system, the policy allows individuals who test positive for AIDS antibodies to remain within the school environment until their attendance is unsafe, either for themselves or others, as indicated by a medical professional. The policy, according to Dr. Ernest Fleishman, superintendent of schools for Greenwich, was drafted after a teacher died of AIDS.

Since the adoption of the policy, there have been no reported cases of AIDS. "Connecticut privacy laws are such that no one is obliged to tell an employer about their medical condition. We never knew that the teacher had AIDS until he died," explains Dr. Fleishman. "The main advantage of the policy, actually the only advantage, is that it has calmed the public by getting the information about AIDS out in the open. Most importantly, the fact that AIDS can not be transmitted by casual contact." Aside from this benefit Dr. Fleishman candidly admitted that "the policy is actually rather senseless. Usually we never even know that someone is infected until it is too late for them to return to the school."

The policy was developed by a committee of district administrators, including Dr. Fleishman and various board members. In addition, local medical directors, most notably Dr. Novak from the Yale School of Medicine, provided medical advice on the disease and its contagious nature.

Although the first of its type, the policy is in no way "earth shattering," concedes Fleishman. The essential nature of the policy provides for confidentiality of any medical data obtained. When a case arises in the school district, a review panel (the superintendent of schools, the head of infectious diseases at the local hospital, and the director of health from the county) examines the situation, and the decision on whether or not to allow the person to remain in the school environment is left in the hands of the superintendent.

The response to the policy, according to Dr. Fleishman, has been mixed—but in general positive. "Of course we had some parents calling us up worried about the child's safety. In those cases we inform the caller that we have gotten the best medical advice available and that we are putting no one in danger of catching the AIDS virus. But for the most part, people have been understanding and supportive."

The case in point: "The teacher who died of AIDS, who precipitated the policy, was recently honored at his school," commented Dr. Fleishman. "A scholarship has been set up in his name, and one of the auditoriums was renamed in his memory."

was the most frequent first response. "Anything that is a potential concern and work disruption to our employees is a concern to us. By easing those concerns, we have a more productive workforce," according to a spokesman from Pacific Bell. "If you don't have a policy, you're continually looking at the down side of the issue," states Nancy Merritt. (The downside might include anything from lingering anxiety to litigation for unlawful discharge.) The company's ability to give early counseling and medical support was the second most frequently mentioned benefit. Moreover, the policy (the paper in the handbook) often came as part of a much larger initiative, a decision to manage health costs, the need to educate, and the like. Policy formation was integral to the mustering of other company forces and programs.

Why single out AIDS for policy formation? The logic is clear: If the prevailing wisdom is to treat AIDS as any other illness or disability, then existing policies and practices apply. No further work is needed. The Bank of America approach of subsuming AIDS under the broader category of "life-threatening illnesses" demands attention here. The practice has been adopted by other organizations to good effect. The same provisions as to leaves of absence, medical cost management, and support through an EAP indeed do apply to cancer, cardiovascular illness, and many other conditions.

But AIDS isn't "just any other illness." No other condition is marked with such stigma and misinformation; no other illness touches the special employee groups discussed in Chapter 1 in the same way; no other illness has the potential for devastation that AIDS has, should spread go unchecked.

WE UNDERSTAND . . . BUT OUR CUSTOMERS DON'T

The public, understandably frightened and confused, and largely ignorant of the manner in which the AIDS virus is transmitted, may interpret any corporate statement as a warning to stay away from the company's products or avoid its premises. Our staff approached this problem from two perspectives: public attitudes, gathered through a specially commissioned public opinion survey to assess the impact on businesses thought to be especially sensitive to public knowledge that an employee was infected; and corporate attitudes, obtained through a series of in-depth interviews with personnel managers in food processing and packaging industries. The results follow.

Restaurants, Hairdressers, and Dentists

A national public opinion poll commissioned by the American Management Association for this report* revealed the degree to which service providers are vulnerable to popular attitudes. According to the poll, conducted in mid-November 1987, restaurants could lose more than half their patrons if it were known that a waiter or cook was infected with the AIDS virus—this despite statements by the Centers for Disease Control and others that AIDS cannot be transmitted via food handling. Barbers and hairdressers would be equally affected, and nearly 60 percent of the 1,250 people interviewed indicated that they would quit their current dentist if they learned that he or she treated patients infected with AIDS.

*Results of this AMA survey are summarized in Appendix B.

While wary of the risk of infection, the public does not believe that employees with AIDS should be fired forthwith. Half of those interviewed believe that a waiter or cook infected with AIDS should be placed on disability leave. An additional 9 percent think that the waiter or cook should be kept on staff, as long as customers are told of the condition—but more (16 percent) would advise the restaurant owner to keep the employee at work and tell no one.

Nevertheless, the loss of patronage which would follow public disclosure that a waiter or cook had the AIDS virus would be devastating to any restaurant. Only 10 percent of those interviewed said they would "definitely" continue to go to the restaurant if they knew a waiter was infected with the AIDS virus, with an additional 28 percent saying they would "probably" continue their patronage. Fewer still would continue to visit a favorite eatery if they knew a cook was so infected. The results were nearly identical when interviewers asked about barbershops and hairdressers.

"Reasonable accommodation" must be reasonable for all concerned, including the customer. In such specialized cases as restaurants—although again, no authority advises that the disease can be transmitted by food—"reasonable accommodation" may mean disability leave, with ongoing pay and health benefits, but without any public acknowledgment of the policy.

Still, these are particulars for relatively narrow and specific circumstances. Authorities agree that most businesses can comply with the surgeon general's policy recommendations without negative result. It is worth noting that every food processor interviewed for this report, if it has taken any stance at all, has decided to treat AIDS within existing policy for illness—but only half of them have publicly announced this decision.

Again, the reader is referred to Appendix B for complete demographics.

Response from the Food-Handling Industry

Our interviews with managers from companies engaged in food processing and packaging fell into three roughly equal groups:

CENTERS FOR DISEASE CONTROL GUIDELINES ON AIDS IN THE WORKPLACE

Summary of Centers for Disease Control Guidelines on AIDS in the Workplace. November 1985

The CDC has recently issued recommendations to provide employers with guidance on the health risks of employing a person with AIDS.

1. The basic recommendation is that an employee with AIDS need not be restricted from work in any area unless they have evidence of other infections or illnesses for which any employee in that area of work should also be restricted.
2. Personal service workers whose services require needles or other instruments that penetrate the skin are urged to follow infection control recommendations that have been issued for health care workers. Instruments that penetrate the skin, e.g., tattooing and acupuncture needles or ear piercing devices, should be used once and disposed of or be thoroughly cleaned and sterilized. Instruments not intended to penetrate the skin, but which may become contaminated with blood (e.g., razors), should be used for only one client and disposed of or thoroughly cleaned and disinfected.

   No special precautions are required for personal service workers whose services do not involve a risk of blood contamination.

3. The CDC does not recommend a prohibition on employment of a person with AIDS working in food services. No evidence exists of transmission of either the AIDS virus or hepatitis virus during the preparation or serving of food or beverages.
4. Workers with AIDS in a setting such as an office, school, factory, or construction site have no known risk of transmitting the infection to co-workers, clients, or customers.

5. The CDC finds the greatest risk of transmission of [HIV] in the health care work place, especially those health care workers who take part in invasive procedures, such as surgery. It is the CDC's position that even health care workers who are known to be infected . . . but who do not perform invasive procedures "need not be restricted from work unless they have evidence of other infection or illness for which [health care workers] should be restricted." [Note: further guidelines applicable to health care workers have been issued.]

1. Companies with written policies. Typical responses:

We established a task force early on, with one person selected to monitor the situation (including public opinion). Subsequently, we developed a written statement, with input from all divisions. Basically, it says, "We treat this as any other illness." The four cases we've had—interestingly enough—were in the service end, rather than direct food handling. We had expected that co-workers would be adverse to working with these persons, and we considered options for moving them. Much to our surprise, exactly the opposite happened. Co-workers were very supportive.

We did a policy clarification and revision—AIDS is a life-threatening illness, treated like any other. At the same time that we announced the policy, we put up posters and distributed literature from the Red Cross. . . . We expected that there might be a public response. There wasn't.

2. Companies following a practice of reasonable accommodation, without having added any specific provisions to existing policy.

We have so many divisions and subsidiaries that we can't do a policy. So, to a great extent, it's just common sense, plus an individual, case-oriented approach. . . . It's education, coolness, fairness, decency, sensitivity . . .that's best.

In practice, we allow the person to work as long as he or she is able. In several cases, we did move the person away from direct food contact. The union agreed to this, and it seemed the best way to handle it.

For many of these respondents, however, operating under "existing policy" meant following guidelines for contagious diseases issued by the FDA. "The same logic and rules apply, even if this is not a food-borne virus. The person wouldn't be laid off; he would be moved from work involving contact with food."

One theme unified many of these responses: the company has an obligation to treat employees fairly and with compassion; in the case of AIDS, however, the company does not have an obligation to tell the world about this practice. Until people are convinced that this is not a food-borne virus, a low profile is best.

3. Companies that are currently wrestling with the problem. Opinions in these companies were divided on what position should be taken. In three cases, the corporate lawyers advised against policy formation, "It's best not to commit ourselves on this" being a typical response.

Perhaps the most interesting finding here is that two-thirds of the respondents have accepted the practice of reasonable accommodation, with or without benefit of specific policy. Opinions differed, however, on the question of whether a worker with HIV infection should be moved away from contact with food. How serious is the issue of business loss? Opinions ranged from "calamitous" to "not an issue—public opinion is changing." The weight, however, shifted toward the latter response.

A COMPOSITE POLICY

In the course of compiling this "composite policy" on AIDS/ARC and other life-threatening illnesses, the American Management Association research staff carefully studied more than a score of existing policies. Our researchers considered both the form as well as the ideas presented. Only those elements which were deemed most suitable and practical for a broad range of companies—big and small, national or local, in the manufacturing or service categories—were extracted for inclusion.

No policy can be installed "turnkey" fashion. For many organizations, policy formation has been the product of task force activity, involving individuals from the employee assistance program, benefits, employee relations, legal, and others. Some have added outside expertise in the form of consultation with medical experts knowledgeable about AIDS. The task force discussions proved helpful in integrating the policy with other plans, including education and benefits clarification.

In addition to the composite policy, we reprint the Bank of America policy on "life threatening illnesses." This document has proved exceptionally valuable in helping organizations think through the policy issues. Both policies are offered to help overcome whatever "writer's blocks" may emerge among those assigned to draft the document.

You will notice that our composite policy tries to minimize the abstract third-party "voice" that is so popular among policy writers. There are two basic reasons for this. First, AIDS is a controversial subject that has elicited highly emotional responses from several segments of society. In the course of discussions with executives, managers, and others who may have to interpret or carry out policy on this subject, we found opinions varied significantly. There was little question that strong personal feelings would "soften" or "harden" any basic policy they would be required to render or relay to employees. So while we normally would expect managers to adjust a broad policy to fit local needs or circumstances, we felt that this policy should be sufficiently plain spoken and direct to allow it to be passed on directly to all employees.

This is a subject that needs understanding, delicate judgment, empathy and sympathy—not always found in the course of normal business activity—and rarely evident in the legalistic abstractions of business and government policies. So we have opted to use a conversational style whenever possible.

You may also notice that Paragraph 1 talks about the appointment of a special coordinator or administrator to deal with AIDS-related matters and, of course, with employees facing other life-threatening illnesses. Most companies have not made this type of appointment, preferring to integrate such concerns into the normal functions of the personnel or benefits departments. HIV-related illness cannot be considered just another insurance matter and assigned to an insurance claims clerk who may not fully appreciate the delicate human and legal aspects as well as the confidentiality of such cases.

THE BANK OF AMERICA POLICY

Assisting Employees with Life-Threatening Illnesses

BankAmerica recognizes that employees with life-threatening illnesses including but not limited to cancer, heart disease, and AIDS may wish to continue to engage in as many of their normal pursuits as their condition allows, including work. As long as these employees are able to meet acceptable performance standards, and medical evidence indicates that their conditions are not a threat to themselves or others, managers should be sensitive to their conditions and ensure that they are treated consistently with other employees. At the same time, BankAmerica has an obligation to provide a safe work environment for all employees and customers. Every precaution should be taken to ensure that an employee's condition does not present a health and/or safety threat to other employees or customers.

Consistent with this concern for employees with life-threatening illnesses, BankAmerica offers the following range of resources available through Personnel Relations:

- Management and employee education and information on terminal illness and specific life-threatening illnesses.
- Referral to agencies and organizations which offer supportive services for life-threatening illnesses.
- Benefit consultation to assist employees in effectively managing health, leave, and other benefits.

Guidelines—When dealing with situations involving employees with life-threatening illnesses, managers should:

1. Remember that an employee's health condition is personal and confidential, and reasonable precautions should be taken to protect information regarding an employee's health condition.

2. Contact Personnel Relations if you believe that you or other employees need information about terminal illness, or a specific life-threatening illness, or if you need further guidance in managing a situation that involves an employee with a life-threatening illness.
3. Contact Personnel Relations if you have any concern about the possible contagious nature of an employee's illness.
4. Contact Personnel Relations to determine if a statement should be obtained from the employees's attending physician that continued presence at work will pose no threat to the employee, co-workers, or customers. BankAmerica reserves the right to require an examination by a medical doctor appointed by the company.
5. If warranted, make reasonable accommodation for employees with life-threatening illnesses who request a transfer and are experiencing undue emotional stress.
6. Make a reasonable attempt to transfer employees with life-threatening illnesses who request a transfer and are experiencing undue emotional stress.
7. Be sensitive and responsive to co-workers' concerns, and emphasize employee education available through Personnel Relations.
8. No special consideration should be given beyond normal transfer requests for employees who feel threatened by a co-worker's life-threatening illness.
9. Be sensitive to the fact that continued employment for an employee with a life-threatening illness may sometimes be therapeutically important in the remission or recovery process, or may help to prolong the employee's life.
10. Employees should be encouraged to seek assistance from established community support groups for medical treatment and counseling services. Information on these can be requested through Personnel Relations or Corporate Health.

Revised 10/25/85

The logical person to act as coordinator is, of course, the director of the employee assistance program. If the organization does not have this function in place, now may be an ideal time to investigate the possibility of adding this resource. There is currently a trend among smaller companies to form consortia or other "shared resources" EAPs with other firms in the immediate area. Alternatively, smaller companies may wish to obtain the services of a medical consultant or specially train a selected personnel manager to act as coordinator.

Lastly, we suggest that your organization's legal counsel be consulted before you implement your policy. Various states, counties, and a few municipalities have passed laws or published guidelines that may impact on the way your company or government agency deals with this subject, and you will need to be continually updated on new medical as well as legal findings. Unfortunately, AIDS is dynamic, a disease that still remains beyond our grasp to control and, eventually, stamp out. But what we do while waiting for a medical miracle—by way of preventive education, understanding, and kindness—will help us all to endure what we cannot cure.

COMPOSITE POLICY

______________________ [name of company] is committed to maintaining a healthy and safe work environment for all employees, as well as providing support for individual employees who may be facing the trauma of a life-threatening or catastrophic illness.

The AIDS epidemic and the spread of infection from the human immunodeficiency virus (HIV) is causing concern in many segments of society. Consequently, some employees may be experiencing anxiety about the possibility of working with a person who has become infected.

The purpose of this policy is to support the physical and emotional health of all employees, minimize disruptions to productivity and morale caused by the presence of a worker with a life-threatening or catastrophic illness, and demonstrate the company's continued commitment to our affirmative action goals related to physically handicapped employees.

As a general principle, the company recognizes that an employee facing a life-threatening or catastrophic illness may wish to continue to work as long as he or she is able. If an individual is able to work, he or she is expected to be productive; if the individual cannot work, then he or she is eligible for health and disability benefits, as specified in [other policies].

As with any handicapping condition, the company will make reasonable accommodations for an employee as long as such accommodations are practical and economically feasible, and in the best interest of the employee and the business unit.

The company reserves the right to ask its appointed physician to examine an employee with a life-threatening or catastrophic illness to determine that this individual is able to work and poses no threat to himself/herself or to others.

The policies and procedures outlined herein apply to all disabilities and do not change any existing medical, benefits, or employee relations policies covering sickness or disability.

Specific Actions

1. Special Coordinator Appointed

To help all employees understand and deal with problems that may arise from AIDS or other life-threatening illnesses, ________________ has been appointed special medical coordinator. He/she has the background, experience, and training to deal with this subject, and can be contacted at [phone number]. The medical coordinator will:

a. Answer all questions that relate to these diseases;
b. Refer employees to proper medical resources, agencies, and organizations that provide tests, treatment, assistance, and support;
c. Discuss assistance and benefits;
d. Consult with the employee, his/her physician and supervisor or manager about any necessary reassignment or adjustment in duties or hours;
e. Circulate the most current information available on this

subject to properly inform employees; and

f. Coordinate and conduct seminars or other programs for co-workers and managers of sections or departments related to AIDS or other life-threatening illnesses.

All employees are encouraged to use the special coordinator as a resource person as needs arise.

2. Confidentiality Assured

In every instance, the special medical coordinator will take every precaution to see that information about an employee's medical condition is kept strictly confidential. Supervisors and managers should also recognize that medical information is personal and confidential and take all reasonable steps to assure confidentiality.

3. AIDS Should Be Reported

Any employee who has reason to believe that he or she has become infected with the human immunodeficiency virus or any other condition that poses a serious threat to health should contact the special coordinator, on a confidential basis.

Any employee who has tested positive for HIV infection or is being treated for AIDS or ARC should report this fact to the special coordinator to ensure that he or she will be eligible to receive support and benefit programs.

A supervisor or manager who learns that an employee has HIV infection or any form of HIV-related illness should counsel that the employee contact the special coordinator. The supervisor or manager may, with the employee's permission, inform the special coordinator of the employee's condition.

4. Employment Ties to Performance

Because HIV infection may take more than five years to seriously affect a person's functional abilities, he or she may be able to work for a long time without any restrictions or problems. As long as an employee is able to perform his or her job properly and meet the standards set for performance, and as long as medical evidence shows that continued employment does not endanger either the individual or co-workers, an employee with HIV infection should and will be allowed to continue working. Employees with HIV infection are entitled to the same working conditions as others and will receive coverage under our various support and benefit programs, as eligible. If, in the course of time, a person with HIV-related illness cannot perform his or her duties, we will make whatever arrangements are necessary to allow that person to work within reasonable limits of his or her capabilities. The person may be assigned to jobs or hours he or she can work. In no case will an employee with HIV infection or any other infectious disease be automatically or summarily discharged.

5. Understanding, Not Ignorance

Employees will be asked to be sensitive to the needs of critically ill colleagues. Continued employment for an employee who is seriously ill may be beneficial, both for personal and for financial reasons.

Next to prompt, professional medical treatment, the most important help that a person with AIDS can get is the understanding and compassion of relatives, friends, and co-workers. In many cases, it is far worse to be shunned by others and lose one's place in society than to suffer the slow progression of disabilities that come from this disease. A policy cannot order anyone to be kind and considerate to the victim of a life-threatening or catastrophic illness. However, we sincerely hope that every employee will do everything in his or her power to make each day a person with AIDS works an affirmation of all that is good and decent in the human spirit.

OTHER MATTERS

Several policies we looked at make provisions for transfer of persons who experience undue stress as a result of working with someone who has AIDS. For example,

> Employees who feel uncomfortable in working with a colleague who is

seriously ill may investigate transfer possibilities on an individual basis, using the company's normal transfer opportunities and practices. No employee will be granted rights to transfer outside the normal transfer procedures.

Such provision, however, has come under criticism at a number of recent AIDS-related business forums and is not included in our composite policy for this reason.

Some consultants who reviewed the composite policy urged that a statement be included regarding the company's position on HIV-antibody testing—the idea being that the company would not engage in preemployment testing. Only one policy we looked at had such a provision, however. The inadvisability of such practices has become a "given." See Chapter 6 and Appendix A.

Many policies we examined contained lengthy statements about the epidemiological and medical facts concerning AIDS and HIV infection—the number of cases, how the virus can and cannot be transmitted, and the like. Although such statements are likely valuable to employees, they seem to fall more under the heading of "education" than "policy." One can argue, however, that such information can and should be transmitted at the time of policy implementation—perhaps in the form of a cover letter. Another common practice is to describe the research that went into the policy formation: the various medical authorities consulted, for example. Along with this, the policy introductions often quote from the surgeon general or from CDC guidelines (see box, page 56), thus lending additional authority to the statement.

4

Planning an Effective Workplace Education Program

Training experts unanimously agree on several points. First, the workplace can be an effective location for AIDS training and education. It reaches groups that might not be reached by other avenues. It may be the only opportunity for adults to put candid questions to an authority, and it creates a "multiplier" effect throughout the organization and into the community.

Second, the experts agree that effectiveness must be judged against the hard criteria of behavior change. For persons practicing at-risk behavior, this means reduction or elimination of that behavior. For persons adverse to working with someone with AIDS, the effectiveness can be measured by the amount of sympathy and support given.

But how to achieve these objectives is the subject of ongoing debate.

The education and training shouldn't be handled by persons who have not been thoroughly trained themselves, say spokespersons from the American Red Cross, the Gay Men's Health Crisis, and others. Not necessarily true, say on-site trainers, who feel they understand

MORGAN MEETS THE CHALLENGE

Stroll through the hushed interiors of the J.P. Morgan Bank, and you're struck by the atmosphere. The edifice bespeaks tradition, old guard values, conservatism. J.P. Morgan represents all these things. And many others.

In the area of employee education on health, on AIDS in particular, Morgan leads the avant garde. An AMA researcher recently attended one of its presentations on AIDS, part of a 1987 theme, "Year of the Body." Considering the hour—8 a.m.—the turnout was sizable. About 30 employees, breakfasting on coffee and danish provided by the bank, settled in to watch a video presentation and ask questions. Dr. William Schneider, Morgan's medical director, Joyce Jenkins, occupational nurse for Morgan, and Public Health Educator Salvatore Licata of the New York City Department of Health co-hosted the program. (More than 300 employees had attended a prior meeting at another uptown location, according to Dr. Schneider.)

The program began with a video, *Sex, Drugs, and AIDS,* produced by O.D.N. Productions and narrated by Rae Dawn Chung, a popular young film star. The video, although it is targeted at teenagers and uses New York City high-school students as actors, had a message for everyone (for more information, see page 81). After the viewing, Licata conducted a question-and-answer session that provoked across-the-board participation. Dr. Schneider wrapped up the program, indicating that the video was available to employees for overnight use.

Among other initiatives, Morgan makes available an AIDS hotline where employees can receive counseling and offers the HIV-antibody test, in strict confidence, free to employees.

Morgan then defied tradition and conservatism by passing out condoms and literature on AIDS. Dr. Schneider summed it up: "Hardly the kind of thing you'd expect at work."

their employees far better than any outside consultant ever could. Moreover, consultants may not have the flexibility to adjust their "packages" to the specific company needs.

Should the education be companywide—or targeted only to specific needs within the organization? The Orange County Task Force recommendations (see Appendix C) favor an all-employee effort, as did many of the early programs discussed in Chapter 1. This is impractical and expensive, said some of our respondents. How does one get to a workforce of 30,000?

Whatever the case, skilled trainers are in short supply. "We get about 20 calls a week for sessions," states Leslie Stein of the Mid-Hudson Valley AIDS Task Force. "Each session we do conduct leads to 10 more requests." Unable to handle the requests with its current staff, the organization plans to mobilize efforts to "train the trainers"—to teach what they know to their in-house counterparts. On a national scale, the American Red Cross is recruiting individuals with medical backgrounds to work as trainers, as well as laypersons who have extensive knowledge about the disease.

Many training consultants are critical of the "one-hour, talking head" approach. (Someone from a local hospital or health department gives a lunch-hour lecture, attendance optional.) "Who's going to be there to mop up after the session?" asks Stein.

Nor has the literature used to disseminate information escaped challenge. A University of Minnesota researcher, Mark Hochhauser, contends that most brochures are written on a level of reading difficulty far above that of the average American. The professor subjected 16 of the most commonly used brochures to standard measurements of reading difficulty, using the Fry Readability Scale. Most of the brochures were written at levels appropriate for college freshmen or above. Most Americans read at about a tenth grade level. The conclusion: the facts simply aren't getting through, the result of what Hochhauser calls "educational malpractice."

Readability aside, others contend that much of what happens in the name of AIDS education simply isn't grounded on an understanding of how adults learn. Adults learn as adults, the information needs to be pragmatic, the explanations clear and straightforward. There's likely to be a great deal of skepticism about anything the government says. Attitudes and habits are deeply ingrained. No small challenge to the trainer.

To get a preliminary reading on how training professionals think through workplace training, we put two questions to a dozen such persons, selecting only respondents who work full-time in such endeavors:

> What questions do you ask yourself before conducting a training session?
>
> What, in your judgment, are the most important keys to an effective workplace program?

The three most frequent answers to the first question:

1. *Where is this audience coming from? What are their needs?* "We need to know what point of view they have; what point of view they expect us to present—ethical, legal, or medical—and whether they want specific information or general," noted one respondent. Another respondent reported that many of the policemen and firemen he had recently trained believe that there was a government plot to conceal the truth about the epidemic. Many had read a popular book expressing this point of view.
2. *What do I really want to accomplish with this session?* That, in turn, leads to other questions. What prompted the request for training? What training—if any—has been conducted prior to this session?
3. *What will happen after we leave?* Is there someone in the company who will determine what needs to be done next?

Responses to the second question were more diverse:

1. *Skilled, well-trained trainers.* In addition to knowing the epidemiological statistics and the medical facts, the facilitator must be able to deal with sexually explicit matters in a way that is appropriate for the audience. In some cases, this means knowing "street language"; in other cases, how to talk about practices that may offend listeners. Moreover, the trainer

cannot have any personal doubts about how the virus is spread. Trainers need to be updated on the progress of the disease. Before presenting any video, the trainer will want to highlight any statistics that have been revised since the taping.

2. *Flexibility in format.* "We learned the hard way," said one trainer. "What plays in San Francisco just won't work in Atlanta." Moreover, the program must be designed to meet specific workplace needs. "Off the shelf" products don't work. The trainer must be able to change plans quickly. Questions and answers—including opportunity for confidential one-on-one discussion after the session—must be planned for.

   As one respondent noted, "If you're not 100 percent sure about an answer to a question, admit it. Tell the person you will find the answer, and either you will call them or they will call you. And by all means, follow up. Don't leave any question unanswered."

3. *Have clear objectives and a way of measuring whether the objectives were accomplished.* See, for example, the before-session/after-session inventory on page 155.
4. *Avoid scare tactics.* The disease is frightening enough. Give information and facts.
5. *Be prepared to deal with special groups*—home attendants, laundry workers, and food handlers, for example. All their concerns can be dealt with under normal provisions for good hygiene and infection control (washing hands, wearing gloves, when appropriate).
6. *Create an atmosphere conducive to learning.* This includes the physical setting as well as a rapport in which participants feel free to ask questions.
7. *Follow-up as part of an ongoing program.* One session won't do it. The training needs to be reiterative and integrated.

In addition, many of our respondents mentioned the need to gain support from upper management; several suggested starting with the management group.

Without doubt, training specialists now have a rich storehouse of materials to draw from in planning and implementing workplace education. Should the organization choose to draw on the services of educators from outside the company, several resources deserve special consideration, including the Gay Men's Health Crisis and the American Red Cross. Should the choice be to plan the program using staff trainers, the workplace education package from the San Francisco AIDS Foundation (brochures, policy planning manuals, videos, and other helps) may be worth examining. In addition, the material presented in Appendix C, excerpted from *Facilitating AIDS Education in the Work Environment,* has been widely noted as an effective planning tool. Many organizations are using a combination of consulting and on-staff expertise.

The remainder of this chapter takes a selective look at some of these resources. Because video presentations provide an efficient way to present basic facts to a wide audience, our discussion includes notes on some of the more popular presentations. The discussion is intended as a representative illustration of what materials are available. No endorsement is intended or implied.

USING THE RESOURCES

Situations may arise that require individual attention, and issues may have to be addressed that are specific to businesses or even to one particular company. We have already seen in Chapter 2 the problems that can arise when an employee's co-workers discover that he or she is HIV positive. Management may wish to solicit outside help in such a case, especially if emotions are running high or the company lacks medical personnel knowledgeable about AIDS. A number of groups will provide speakers and counselors to talk with employees.

One such group is the Gay Men's Health Crisis (GMHC) in New York City. The GMHC, which is nonprofit and does not charge for its services, employs a staff of 25 in its educational and training department and has access to nearly 100 volunteers in its speaker's bureau, many of whom have professional credentials. Before going through what the GMHC terms a "Health Belief Model," a five-step approach is designed to create an attitude/behavior change. After an introduction, the speaker uses questions from the audience to evaluate their preplanned needs assessment, and to decide on language appropriateness. According to Jim Holmes, coordinator for information services, "It is important to know the audience and to find out if what you perceived, prior to meeting them, is accurate. Knowing how to talk to them, and giving them information that reflects their attitudes, are important elements of any education program."

Once the speaker has identified the audience, he or she needs to begin to break away at the rubble, the misconceptions surrounding AIDS and HIV infection. "A piece of literature, or a lunch-hour meeting, is not going to accomplish anything," believes Holmes. What will is a thorough program designed to help people understand the virus and the emotions connected with it. The GMHC attempts to "put a face on the epidemic" and reach its audience with information, not fear.

The Five Steps

1. *Help the client recognize his or her own vulnerability to the infection.* According to the GMHC, defenses are high when dealing with AIDS and HIV infection, with the tendency to divide the workplace and the world into "them" and "us." The idea during this step is to make the people in the audience realize that HIV can infect anyone, and sooner or later someone they know (be it a neighbor, friend, family member, or co-worker) will contract the virus.

   After the audience has accepted the reality of the situation, the speaker has the option of bringing in a person with AIDS to talk to the audience and answer questions. The importance: a person with AIDS is still a person.

   This tends to be the "down" point of the program. It is important, from then on, to make everything more positive, to deal with the ability to live with the virus, as opposed to dying from it.
2. *Discuss the preventability of infection.* During this step, the speaker can talk about what is really known, medically, about

HIV infection. Evidence is available about the transmission of the virus; thus, the speaker can talk about how not to become infected.

3. *Tell them that anybody can make the behavior changes that will protect them from HIV infection.* This puts control in the hands of the individual. People are afraid because they think they have no control over preventing infection. The statement becomes: I am not going to become infected with HIV because of a decision somebody else made. It is in my control.
4. *Discuss HIV infection as a societal issue.* This means lending support to people with AIDS and those who are HIV infected. This involves talking about sex and sexuality, not as something forbidden, but natural. Once one accepts the nature of the transmission, one can accept those who are infected.
5. *Point out that people can still live fulfilling lives after making behavioral changes to contain the spread of HIV infection.* What this means is being responsible for both yourself and the people around you. According to Holmes, many people believe that they have to give up a lot to protect themselves and others from HIV infection. That's not the case. What they do give up is irresponsibility. What they gain is a sense of control, a control that stems from knowledge. From a workplace viewpoint it becomes a positive giveback to employees—in terms of knowledge and policy guidelines for their protection.

According to Holmes, "Once you take someone through the five steps it no longer becomes a question about whether they are going to make a rational decision in policy. It becomes a human decision."

THE AMERICAN RED CROSS—A CHRONOLOGY

A Horror Story. The news breaks that people who received blood transfusions prior to 1985 may have received blood contaminated with the HIV virus. Just as the public was beginning to learn about HIV infection, people who were previously excluded from the "high risk groups" were beginning to develop the virus.

Known for decades as the resource for helping the sick and providing blood to hospitals and clinics around the world, the American Red Cross faced a formidable challenge. And to further complicate the matter, at risk were not only transfusion recipients, but also hemophiliacs, whose clotting agent is developed from blood by-products. By only one HIV-infected donor supplying blood to that mixture, thousands could be infected. The magnitude is frightening.

A test to detect HIV antibodies would not be found until 1985 (note that then, as now, the test is not 100 percent accurate). Thus, in 1983, when the HIV virus was identified, the Red Cross began educating current and potential donors about the disease. With no cure or vaccine on the horizon, careful screening of donors was essential.

. . . And It Continues . . .

In October of 1985, the Red Cross turned its education plan toward the general public, making a long-term commitment to educate everyone about HIV infection, with the primary emphasis on prevention.

What needs to be remembered is that infection through blood transfusion is only a small percentage of the problem. Education of the entire population to influence lifestyle change is the real issue. Since 1985, the American Red Cross has been vigorous in this pursuit.

With the assistance of the Centers for Disease Control (CDC), the American Council of Life Insurance (ACLI), the Health Insurance Association of America (HIAA), the Public Health Service (PHS), and various other organizations, the Red Cross developed and distributed AIDS public education materials.

The Contribution:

- Since November 1985, 36 million brochures with information on AIDS have been distributed, in conjunction with the PHS.
- In 1986, in association with the ACLI and the HIAA, the Red Cross produced the award-winning documentary *Beyond Fear*. Airing on the American Forces Information Service network,

among other channels, it is also used by the American Medical Association to train physicians in AIDS education and by the New York State School Board for viewing in 750 school districts around the state. The video is currently being produced in Spanish; it should be available in early 1988.

- In June 1986, with the help of the PHS, the insurance industry, and the Advertising Council, the Red Cross began its AIDS Public Service Advertising Campaign, providing information on the three major media vehicles—television, radio, and print. The campaign, with its slogan, "Rumors are spreading faster than AIDS," addresses the bottom line, by providing the public with facts about casual contact, and the real ways you can be infected.
- To date, 600 Red Cross centers nationwide have designated special AIDS coordinators to implement AIDS information and education programs.
- With the insurance industry, five million copies of "AIDS: The Facts," a generic brochure, have been distributed—in both English and Spanish.
- In addition, the Red Cross has worked with organizations representing various minority groups, including the Urban League and the Labor Council on Latin America, to reach their constituencies—with the film *Beyond Fear* and other collateral material (some of this material has been translated into Spanish, and more is being produced).
- Youth programs. In 1986, the Red Cross made a concerted effort to reach the teenage population (predicted to be the next largest section of Americans to be infected by the virus). Endorsed by the National Association of State Boards of Education and the National Institute on Drug Abuse, the video *A Letter to Brian* addresses the transmission of AIDS through drug and sexual activity. Accompanying the video are student workbooks, a teacher's guide, and parent support brochures. It is available for both school use, and on a free home loan to parents.
- The workplace. In November 1987, the Red Cross released three vignettes to be shown after the completion of *Beyond Fear,* addressing specific situations in the workplace. The vignettes,

titled *Working Beyond Fear,* raise questions for the purpose of promoting discussion and are only shown as part of a seminar given by Red Cross counselors.

In addition to these programs, the Red Cross offers training and services for people with AIDS and those who take care of them. These include transportation for AIDS patients to doctors, churches, and other social service activities, as well as training. The training for nonprofessional caregivers is an 18-hour course on caring for AIDS patients in the home. For professional home attendants, at a minimal fee of $10, the Red Cross conducts an 8-hour training course specifically on AIDS care.

THE VIDEOS

The American Red Cross and many other information and health resource companies have produced videos aimed at the issue of AIDS. For the most part, they are short films containing general information for a wide range of audiences and are designed to open discussions on the AIDS topic led by health educators. *Caution:* because of the evolving nature of the virus, information in the video tapes may need to be updated; therefore it is vital that a expert on AIDS be in attendance whenever the films are shown. The following list is selective and by no means includes all the films available.

To confront the AIDS issue, the Red Cross has developed three videos aimed at the general public, children, and the workplace.

The most basic of these productions is *Spread Facts Not Fear.* This general audience film provides basic information about the virus, including description on how it works, transmission, misconceptions, and blood donation.

Discussed in a simple, straightforward manner, the video's message is this: we must take personal responsibility in order to prevent infection. The fears we manifest in connection with the infection have to be faced in order to understand, confront, and eventually overcome the virus. (Running time: 11 minutes)

The second film, *A Letter to Brian,* addresses the issue of transmission among the teenage population. Teenagers represent the group that is most open to experimentation with drugs and sex. This, combined with the perception of immortality, creates an increased risk, according to the video. Filmed as a dramatization, and narrated by Michael Warren (formerly of Hill Street Blues), the film discusses the transmission of the disease and its effect on the body. The video contains a question-and-answer segment with Dr. C. Everett Koop, in which students address questions to the surgeon general.

The Red Cross reports that the film has been frequently used in workplace settings, because it addresses parents, a substantial portion of a company's employee group. Although the presentation is simple, the film introduces some very serious questions—making it important that an adviser be on hand to provide answers, according to a Red Cross spokesperson.

The central message: AIDS is everyone's concern, and you can never be sure who has it. Released in November 1987, the video comes with an instructor's guide, student workbook, and parent's brochure. The video is available through your local chapter of the American Red Cross and is also available for the classroom and the home. (Running time: 29 minutes)

Intelligent understanding, coping and prevention: these are the keys to the American Red Cross's most comprehensive video on the AIDS epidemic, *Beyond Fear*. Available in two lengths, one hour or 30 minutes, the video attempts to address the misconceptions concerning the virus by presenting the facts, not the headlines.

The film is narrated by Robert Vaughn. Various doctors and health care specialists present information on the disease, from discovery to infection, spread, symptoms, and prevention. These professionals represent such organizations as the Centers for Disease Control, the National Institute of Health, the Washington Business Group on Health, and the Gay Men's Health Crisis.

At the core of the video is the information that AIDS cannot be spread through casual contact. Interviews with people who have AIDS, and with their friends and families, address this point. These claims are further substantiated with medical information.

Produced in association with the Health Insurance Association of America, the 30 minute version is available from local American Red

Cross chapters. The hour-long version is available by contacting Modern Talking Picture Service, St. Petersburg, Florida. According to one of the counselors at the Red Cross, the shorter version is better suited for most audiences because of the nature of the information: "The audience becomes 'antsy' when confronted with the issue over a long period."

In what they term "AIDS 101," in combination with *Beyond Fear,* the Red Cross will provide a counselor to come to the workplace, discuss the changes in medical information to date, and conduct a question-and-answer forum with the group. If the group or company has a medical director who wants to conduct the meeting, a Red Cross adviser will preview the film with that person to acquaint him or her with updated information and answer any questions or concerns prior to the group meeting.

To confront directly the issue of AIDS in the workplace, the Red Cross has produced a series of vignettes to be used as a companion piece to *Beyond Fear.* The vignettes, titled *Working Beyond Fear,* present three separate dramatized situations, referencing three aspects of the issue. The first one addresses a worker's right to privacy after confiding in a co-worker his positive test for the AIDS antibody. The second looks at a situation revolving around ARC among heterosexuals, and the third addresses the issue of casual contact when an employee's family member dies of AIDS.

Each vignette looks at the situations in a realistic manner, spotlighting questions that are, unfortunately, very real:

1. "What's a little blood among friends?"
2. "We don't have, you know, kinky sex."
3. "What happens if your son or daughter gets AIDS? Do we just get rid of you?"

The purpose of the vignettes is to open up discussion in the workplace, and to answer many questions that have previously been left unanswered, as well as to open up new areas. The *Working Beyond Fear* vignettes are available only as part of a two-hour seminar program, conducted by American Red Cross counselors; the program includes the video *Beyond Fear.*

With all the American Red Cross videos, up-to-date information on the virus should be provided prior to the showing. The nature of the disease is such that new information is constantly arising—leaving some statistics dated and misleading.

Rental for *Spread Facts Not Fear, A Letter to Brian,* and *Beyond Fear* is $10 per day, or you can purchase the cassettes for $45. It is available in 3/4 and 1/2 videocassettes. The seminar costs $175 for 20 to 30 participants.

AIDS: Changing the Rules, put out by AIDSFILMS, is a grown-up version of its teenage counterpart, *Sex, Drugs and AIDS* (reviewed on page 81). Hosted by Ron Reagan, Beverly Johnson, and Ruben Blades, this film addresses hetereosexuals and tells them to protect themselves; AIDS is no longer a disease of homosexuals and IV drug users only. Like *Sex, Drugs and AIDS,* it pulls no punches: AIDS patients talk to the audience; the pictures showing Kaposi's sarcoma symptoms are not pretty. Still, the film finds room for some humor to relieve the tension. Ruben Blades talks about how to use a condom. Three women discuss using condoms (almost a carbon copy of a similar scene in *Sex, Drugs and AIDS*), and a heterosexual couple debates using one before they make love. Running time: 26:10 minutes.

AIDS: What Everyone Needs to Know, distributed by Churchill Films in Los Angeles, is comprehensive in its scope and presents the facts in a straightforward, no-nonsense way. It explains precisely what AIDS is and how it interferes with the body's immune system, and it identifies risk groups. Most important, it clearly specifies the activities and mediums that transmit the AIDS virus and those that don't. Dr. Michael Gottleib of the UCLA School of Medicine, who identified the AIDS virus in 1981, and Wendy Arnold, a health educator with the AIDS Project Los Angeles, both appear in the film and acted as consultants. Also appearing are a number of AIDS patients who describe their symptoms and add a human element to the story. The video comes with a booklet of suggestions for discussion leaders, including a partial listing of resources for additional information. Running time: 19:30 minutes.

One may wonder why the Creative Media Group of Charlottesville, Virginia, decided to use cartoon characters to present its film,

AIDS Alert. "Dr. Good Health" and his audience, in the form of gender symbols with faces drawn onto them, hold a question-and-answer session meant to address people's most common fears about AIDS. Meanwhile, the AIDS virus itself is illustrated by a squat, mean looking goblin complete with mustache and cap who resembles a New York City mugger. Accompanied by sound effects, the AIDS goblin is shown beating up the immune system, diving into and swimming around the bloodstream, and whistling over his evil buddies, Mr. Kaposi's sarcoma and Mr. Pneumocystis (meant to represent the opportunistic infections associated with AIDS). Dr. Richard P. Keeling, Director of the AIDS Task Force for the American College Health Association, appears at the beginning and end of the film to reassure us that although we should be concerned about AIDS, we can protect ourselves and control our risk. Comes with a "Discussion Leader's Guide." Running time: 22 minutes.

Sex, Drugs and AIDS, put out by O.D.N. Productions, is designed to discuss AIDS education with teenagers. Rae Dawn Chong is host narrator, and she talks frankly to teenagers on their own level and in their own language about AIDS. Basically she tells the kids three things: (1) AIDS is hard to get so you don't have to worry about casual contact; (2) you can get AIDS by sharing needles, so don't shoot up; and (3) you can get AIDS by having sex with someone who has the virus, so use a condom or don't have sex at all. The film uses some very powerful means to get its message across. It shows two kids actually shooting up and injecting themselves with each other's blood. Five real-life AIDS patients representing the five risk groups tell how they each got the disease. A bike store owner talks about his kid brother who died of AIDS and how the experience changed his attitude toward homosexuals. Three girlfriends talk intimately about having sex: one tries to persuade the other to use condoms, while the third decides not to have sex at all. A whole host of normal activities, such as kids sharing pizza or shaking hands, are shown to demonstrate that AIDS cannot be transmitted by casual contact. The only possible drawback to this film is that it is too simplistic: it leaves out some things and kids will inevitably ask questions. It should not be shown, therefore, without supervision. It comes with a discussion guide. Companies are showing the film to employees with the understanding that the video may be borrowed for viewing at home. In this way management shows that it cares about its workers, their families, and their community.

An Epidemic of Fear—AIDS in the Workplace, financed by Pacific Bell, Bank of America, Chevron, and Levi Strauss, confronts the issue of AIDS in the workplace straight on. Interviews with employees with AIDS, their co-workers, and medical experts look at the issue of casual contact, as well as the psychological impact of the virus. These interviews take place in the workplace and in the homes and hospital rooms of the employees with AIDS. The essential message is that employees with AIDS need to feel a part of the group, to be productive and in control of their lives. An interview with a man with AIDS, who looks healthy, but has had both his job and insurance terminated because of his condition, highlights the alienation many of these individuals feel from the world. The video has been donated to the San Francisco AIDS Foundation, to be used for fund raising purposes. The video may be purchased alone, or as part of a five-part comprehensive education program titled *AIDS in the Workplace.* This includes:

- An educational guide for managers
- Strategy manual
- Strategy manual appendix
- Brochures

One of Our Own, produced by Dartnell Corp., in Chicago, blends corporate lessons on how to deal with AIDS with human drama.

Jeremy Weldon, president of Weldon Associates, a Chicago-based ad agency, has an opportunity to capture the MasterData account. The stakes are high, and he needs his best talent on the assignment. But award winning copywriter Tom Bramson has been off-speed recently. Jack Hadley, vice president and creative director, has some persuading to do before Weldon will agree to put Tom on the team.

Late that afternoon, Bramson confides in Hadley. "Remember when I was worn out last spring? . . . Well, last month I found out it was more than that. I was going through something called ARC . . . They told me that the ARC phase is over. What I've got now is AIDS."

The scenes that follow typify stories occurring in thousands of offices nationwide. Bramson wants to keep working; copywriting is his life's work, and he still has a contribution to make. Moreover, he wants his condition known to his co-workers. Human Resources Director Kathy Anderson knows what to do, but the education program she launches doesn't take hold. Tom becomes a pariah in his own office. Lunches are canceled. A production manager quits. Tom comes up with a winning slogan for MasterData, but account executive Bill Sharkey won't let him handle the story boards.

Abbie White, the bookkeeper, reaches out. By taking a sip of coffee from Tom's cup, she shows that she understands. Her disclosure: her son died of AIDS. He died alone, and she doesn't what that to happen to Tom. Gradually, other employees work through their fears to confront the human tragedy of "one of their own" dying from the disease.

The lessons are clear but unobtrusive: persons living with AIDS can make a contribution. Employees need *facts* and *time* to come to terms with deeply ingrained fears. Support from the CEO is a key component. On the legal and ethical issues, the script has Weldon Associates holding a federal contract, leaving no doubt that Tom is protected as a handicapped employee. "Even if we didn't have a federal contract," says Hadley, "I hope the agency would do what's ethically right and keep Tom on."

The video comes with a 36-page meeting leader's guide. Also available are a manager's guide and two brochures. One of the brochures is targeted to employees who find themselves working with a person with AIDS. Running time: 30 minutes.

OTHER SOURCES

The Workplace Health Communications Corporation, in Albany, New York, has put together a number of services and products addressing the AIDS issue. It begins with *All About AIDS,* a bi-monthly newsletter updating key AIDS topics. Designed for everyone inside and outside of the organization, this newsletter includes legal and medical changes and organizational vignettes designed to reduce anxiety and confront legal issues.

A 32-page booklet, titled *Caring About AIDS,* is designed for managers and employees and gives basic information about AIDS transmission, how to stop the spread of AIDS, how to work with people who have AIDS, and the legal issues.

AIDS Training Tools for Employees: Slides and Guidebook is a complete set of interactive training media to be used by trainers and human resource managers to train employees about the medical, legal, and psychological issues of AIDS. It includes five separate slide shows and audio cassettes: *Medical Facts, Demystifying AIDS, AIDS and the Law, Preventing AIDS Transmission,* and *Working with People with AIDS.* The guidebook accompanies the slide shows and also includes training exercises, questions and answers, glossary, and referral sources.

Managing AIDS in the Workplace: An Executive Briefing and Training Manual is a 100-page looseleaf manual with chapters on the medical facts, the economics of AIDS, education, testing and confidentiality, employee and public relations, current case law, CDC guidelines, case studies of corporate AIDS policies, and referral sources.

The program was produced in cooperation with the Institute for Disease Prevention in the Workplace. The latter, a nonprofit organization, also provides custom training and consulting services on AIDS, nationwide.

PAMPHLETS AVAILABLE FROM THE AMERICAN RED CROSS

1. AIDS: The Facts.
2. AIDS: What You Should Know.
3. AIDS and Children.
4. If Your Test for Antibody to the AIDS Virus Is Positive.
5. AIDS and Your Job—Are There Risks?
6. Caring for the AIDS Patient at Home.
7. AIDS, Sex, and You. General Information.
8. AIDS and the Safety of the Nation's Blood Supply.
9. Facts About AIDS and Drug Abuse.
10. What Every Parent Should Know About AIDS.

PAMPHLETS FROM THE GMHC

Packs:

1. Info Pack (contains items marked #, English version only)
2. Gay Pack (Info Pack plus items marked *, English version only)
3. Med Pack (contains items marked +, English version only)

Brochures:

1. AIDS Hotline(#) (bi-lingual—English/Spanish, also in Chinese)
2. When a Friend Has AIDS(#) (available in English or Spanish)
3. Women Need to Know About AIDS(#) (available in English or Spanish)
4. I Can't Cope with My Fear of AIDS(#)
5. Condom Guide for Men and Women(#)
6. Ounce of Prevention(*) (bi-lingual—English/Spanish)
7. Safer Sex Guidelines(*) (street language, available in English or Spanish)

Booklets:

1. Medical Answers About AIDS(+)
2. Legal Answers About AIDS
3. GMHC Client Services Directory

White Papers:

1. Drugs and Their Side Effects
2. Overview of Psycho-social Issues(+) (available in English or Spanish)
3. Infection Precautions for PWAs(+) (available in English or Spanish)

RESOURCE LISTING

Gay Men's Health Crisis, Inc.
Publications Orders
Box 274, 132 West 24th Street
New York, New York 10011
(212) 807-7517

O.D.N. Productions
Sex, Drugs and AIDS
74 Varick Street, #304
New York, New York 10013
(212) 431-8923
price: $335 to buy,
$85 to rent (per week)

Creative Media
Health Alert Division
AIDS Alert
123 Fourth Street, N.W.
Charlottesville, Virginia 22901
(800) 255-3517, ext. 50

AIDSFILMS
AIDS: Changing the Rules
50 West 34th Street
Suite 6B6
New York, New York 10001
(212) 629-6288
price: $40 to buy, no rentals.

Modern Talking Picture Service
Beyond Fear (one-hour version)
5000 Park Street North
St. Petersburg, Florida 33709
((813)541-5736
free-loan
(additional videos available)

San Francisco AIDS Foundation
An Epidemic of Fear:
AIDS in the Workplace
333 Valencia Street
P.O. Box 6182
San Francisco, California
94101-6182
(415) 861-3397
video: $275.00
total package: $398.00

The Workplace Health
Communications Corporation
4 Madison Place
Albany, New York 12202
800-334-4911
800-942-1002 (in New York)

Orange County Task Force
c/o Pacific Mutual Life
Insurance Company
700 Newport Center Drive
Newport Beach, California 92660
714-640-3014

American Red Cross
Contact your local chapter

Churchill Films
AIDS: What Everyone Needs
to Know
662 North Robertson Boulevard
Los Angeles, California 90069
213-657-5110
800-334-7830
price: $275 to buy,
$60 to rent (3 day)

Dartnell Corporation
One of Our Own
4660 Ravenswood Avenue
Chicago, Illinois 60640
312-560-4000
Also distributed by the American
Management Association
(see inside back cover)

5

Case Management

Hard questions occasionally have simple answers. With the AIDS issue, we face a problem of enormous complexity—a disease that destroys its victims' lives on all fronts: physical, emotional, spiritual, and financial. Such total bankruptcy of human resources by a single microorganism is unprecedented in modern medicine, and solutions seem far off. However, recently an answer that can at least partially remedy the financial dilemma of AIDS has come within reach.

Case management in the care of AIDS and other catastrophic illnesses has offered substantial savings to those employers implementing it. "It's pretty much an 'everyone wins' philosophy," says Jan Roughan, former business manager/director for Equitable Assurance Society's Case Management Services. It seeks to provide the most appropriate care without the constraints imposed by most benefit plans. Briefly, the approach facilitates the best use of existing

---

The editors wish to thank Joanne Hilferty of The Health Data Institute for reviewing this chapter and her organization for providing case study and interview material.

benefits and makes home care, hospice care, and other outpatient benefits available that may not be covered, resulting in major savings to the employer and more humane care for the person with AIDS.

For many AIDS patients, home care is by far the preferred approach. "I know from personal experience that home care is more cost effective—and just simply more humane," stated Gerard Wagner of Bantam, Connecticut, in AMA's 1985 study. "I recently cared for a terminally ill patient at my home. The three weeks of care cost $1,200. The same care in a hospital would have cost $21,000. During my friend's illness, we received wonderful support from a local hospice and from the service people who delivered our supplies."

The savings in this instance was almost twenty-fold, the care received, optimal.

Case management promises dramatic savings in expenditures for all long-term and catastrophic illnesses. But the approach is new: cost savings are long term rather than immediate; benefits managers are unfamiliar with the practices and the terminology; effective implementation demands competent medical expertise on the part of the case manager and in the utilization review (UR) procedures often used to support such services. There are wide variations in what is presented as utilization review and case management. Therefore, the purchaser should be careful to define what is included in the particular program. In addition, some authorities have raised the issue of liability for the company (if self-insured) or for the carrier, should decisions be made to the detriment of the patient.

In spite of these pitfalls—real or imagined—many in the industry believe that the potential more than justifies whatever effort may be needed to overcome the problems.

CASE STUDY

The following case study, courtesy of The Health Data Institute, documents the care of an AIDS patient. The case involves three separate hospitalizations and illustrates the savings generated by a case management approach.

The patient was a male, age 46. The first hospitalization, on March 11, 1986, for bilateral pneumonia, occurred prior to the involvement of a case manager. The case was referred to case management services, however, at a time that was opportune for exploring the support systems available to him. His case manager and doctor suggested that a local hospice was appropriate and his case manager noted that his disability benefits would run out the following November.

The second admission, on April 15, 1986, occurred as a result of an acute psychotic episode, just a week after the patient had been discharged to home. A medical assessment of his condition showed some brain atrophy, although central nervous system infection was ruled out. The patient was exremely weak and unable to take care of himself.

After five days in the hospital, he no longer needed an acute care setting; however, because of his deteriorating physical and mental status, he could not be left alone. His physician was willing to discharge the patient provided that a 24-hour attendant be called in to care for him at home and a licensed practical nurse be available daily to administer medication. Home care for the patient involved an exception to benefits, which was readily arranged by the case manager. The case manager contacted a home care agency and arranged for 24-hour personal care attendants and a licensed practical nurse (LPN) to administer his medication.

The patient's discharge plan also included regular visits from hospice volunteers and the local AIDS Action Task Force. When the case manager and LPN agreed the patient could safely self administer his own medications, the LPN's services were discontinued. The patient's condition continued to deteriorate and he became irrational, ordering his attendant to leave his apartment. But because of the contact established with community resources by the case management team, he was not left alone for the two days during which he refused care. The volunteers from the local AIDS task force and from the hospice service continued to care for the patient. After two days, he allowed his attendant to return, but, as his condition continued to deteriorate rapidly, he was readmitted to the hospital shortly thereafter. He died on May 19.

The final cost savings was as follows:

Days saved:	27 days private room @ $525 × 2.5 for estimated ancillary charges	$35,437
Substituted care:	Personal care attendants 24 hours a day × 20 days	(3,600)
	LPN 4 times per day × 7 days	(450)
Costs saved:		$31,387

Carol Anne Delany, R.N, who was the project manager, evaluated the outcome of case management intervention in this instance as follows:

> ______________________ was home for a period of 27 days. Early intervention in this case and the case management process allowed this patient in the terminal stages of his disease to remain in the familiar and supportive environment of his home. The exception to benefits for home care and arrangements initiated by the case manager for additional support through friends and volunteers from the AIDS Action Task Force and the ______________ Hospital Hospice Program resulted in a substantial cost savings and quality care.

As this case study suggests, an incident of AIDS need not wreak financial havoc with an employee benefits plan. These costs *can* be managed. Paul Gertman, M.D., formerly chief scientist of the The Health Data Institute, concurs: "AIDS is a tough problem, but major employers can handle many aspects of it." He outlines three key actions:

1. Commitment by the CEO and top management to "manage" AIDS issues.
2. Realization that prevention is crucial. Don't wait until your company is in trouble.
   - Develop policies and procedures in advance
   - Train key staff
   - Designate a coordinator to manage a crisis if one occurs
3. Have a medical case management program:
   - To ensure that individuals get good care by qualified specialists

- To ensure that care is provided cost effectively

Smaller companies and those that are self-insured are at greater risk. The self-insured organizations must, of course, be sure that their stop-loss coverage is adequate. But case management can be very effective for such companies, say Paul Gertman, M.D., and others. Smaller firms may, however, have to wait longer for a payback on the case management investment (the smaller the workforce, the less the likelihood of major illnesses within the first years of implementation).

DEFINING THE TERMS

The terms "case management" and "utilization review" are new in the benefits management vocabulary. The definitions have yet to solidify. In addressing AMA's annual Compensation and Benefits Conference, Dr. Anthony Gadja, health care economist, summed up the confusion surrounding the definition of utilization review: "[It] means different things to different people." Concurring with this, Joanne Hilferty of The Health Data Institute says, "One of the problems in the field is that each of the terms [managed care, case management, utilization review] is being used in ten different ways. We all have to be careful to define our terms before we speak."

The case study presented above and the box on pages 93 to 95 offer working definitions of case management in the narrow sense of the term: A case manager is assigned to work with the patient, the doctor, and the family in facilitating the best possible care.

This, of necessity, must involve some form of utilization review: an examination of the type of care that is proposed or conducted.

Thus, the terms describe overlapping services, and some practitioners are broadening the term "utilization review" to include case management. For purposes of this discussion, however, we have chosen to use the terms in their narrowest meanings. This helps focus more closely on the individual components of the services involved. Moreover, it avoids confusion, in that not all utilization review programs may make provisions for assignment of a case worker (case management in the narrow sense).

WHAT IS CASE MANAGEMENT?

In 1985, we spoke at length with Jan Roughan of Equitable. Her definition of case management describes the heart of most case management programs:

First, let me give you an example of how the case management system works and then provide some background on the concept behind it.

One case involved a two-year-old child who received a contaminated blood transfusion as an infant. She had been hospitalized several times for treatment of what we now know were AIDS-related complex (ARC) symptomology and, later, "opportunistic" diseases. The definitive diagnosis of AIDS was made during a hospital stay in August 1985. Case Management Services became involved to assist the physician and mother (a single parent) in implementing a long-term treatment program for the child.

What we did was talk to the mother and the physician. The physician said that the child really could be cared for at home. But we had a working parent here, and so what we did was institute an in-home program, with all the necessary supplies and services, so that the mother could stay at work and keep her health coverage for the child.

So often with cases like this, when home care coverage is denied, the child is ultimately discharged from the hospital with little or no preparation for a long-term plan of care. The mother is the only person available to take care of him or her, so she stays home, loses her job, goes on welfare—you know the story. It is a story that has very little logic—from a human, a medical, or a cost effective standpoint.

Case management is a program designed to facilitate funding based on the medical logic of a situation. The concept actually is an extrapolation from the workers' compensation and long-term disability rehabilitation market. We merely took it a step further and said it makes sense to match funds with the logic of the situation—a point that I can't emphasize strongly enough.

We just recently added terminal cancer and AIDS to the list of injuries and illnesses covered. Previously, the program applied to head injury, spinal cord injury, multiple fractures, severe burns, severe strokes, high-risk infancy, amputations, multiple sclerosis, and amyotropic lateral sclerosis (better known as Lou Gehrig's Disease).

Often with catastrophic illness or injury, the appropriate care is not prescribed—because there are constraints built into the policy language and there is no mechanism for funding based on medical logic. Take, for example, a spinal cord injury. Many local or community hospitals simply do not have the mechanisms or the expertise to deal with the long- and short-term needs of the patient. Under the case management system, we would redirect the patient to a regional spinal cord injury center, of which there are 17 in the United States, or to a facility we know has developed the level of expertise to deal properly with this injury. We would then work with the family and physician to facilitate discharge from the acute care facility, and put the person in touch with programs that would expedite reentry into society. We would allocate funds to modify the home where necessary to provide a barrier-free environment. There might, for example, be a need for larger doorways, for a grab bar in the shower, and the like.

This approach prevents a long hospitalization, a dependence on hospital care, and a mindset that inhibits reentry. The approach is not only care effective but also cost effective.

We say that the program philosophy is to facilitate an optimum level of recovery, and we mean just that. We don't call it rehabilitation because so many people think that rehabilitation means putting Humpty Dumpty back together again.

We want to effect a timely discharge and put the necessary services, support, and supplies in place for an optimum level of recovery—whatever that may be in a particular case. At the same time, we offer psychosocial support for the family. Each case is assigned to a coordinator, a health care professional who's seasoned in orchestrating care for the catastrophically ill or injured. This person works with the claimant, the family, and the physician as an added team member to plot out the plan of care and to make sure the resources and finances are there to make the plan successful.

Our companies are moving in the case management direction, and several of these are in the start-up phase. It's a prudent move. In today's market, if a carrier doesn't have a cost containment program (of which case management is one example), that company isn't competitive.

Other companies have studied our approach to case management for two years and are following suit. We want to see them succeed as this affords additional substantiation for this dearly needed service. In some ways, it's an atypical product for an insurance carrier—and it's a very sophisticated kind of endeavor.

There's clearly a trend in the industry toward greater coverage for home care. For many companies, the difficulty lies in putting a mechanism in place that will allow them to base their allocation of funds on the claimant's short- and long-term needs. Often the mechanism does not exist for coverage of home care at the same reimbursement level as for hospitalization. If it's a fully insured account, ERISA guidelines dictate that you treat each and every individual and each individual's dependents exactly the same. You can't make exceptions for one and not for another.

This, however, has been the approach in many instances; companies find themselves in a precarious position by allocating funds on an "administrative exception" basis. But the case management system is added as a special benefit. It's not that the other carriers don't see the efficacy of home care. It's that the other carriers don't have the mechanism in place. Others are looking only at the short term, denying anything that's not provided for in the policy. In the longer term, this approach is probably going to cost them a lot more.

The remainder of this chapter is organized as follows:

First, a step-by-step discussion of how case management works, with some form of utilization review in place.

Next, a close-up on what spokespersons in the insurance industry have identified as the "critical success factors" in implementing case management: communication within the company, the background of the case manager, and the all-important role of utilization review.

Finally, a look at some broad actuarial guidelines potentially useful in anticipating the cost of AIDS.

THE CASE MANAGEMENT PROCESS

Case management typically involves three steps: assessment, coordinating and planning, and monitoring treatment.

Step 1. Assessment

For case management to be most effective, early intervention is crucial. In the case of AIDS and other illnesses, the sooner an employee or his or her family member is identified as having the disease, the greater the potential for cost savings. From a medical standpoint, rapid diagnosis of AIDS offers the best opportunity for managing the case humanely and prolonging life.

Pinpointing situations that lend themselves to case management requires ongoing review of a company's cases. This may be done out-of-house by an independent review company or a carrier, or in-house by a trained professional who conducts case management as part of the benefit plan administration.

"Trigger points" or "alert" mechanisms provide a signal that a particular claim should be referred for case management. The point at which the alert is triggered can vary widely. Probably the earliest possible intervention can be achieved by using precertification, or preadmission review. Indeed, Honeywell's program is actually a step ahead. At Honeywell, health care professionals, usually occupational nurses, may be called on to review a health risk assessment and in doing so may spot employees who may be at risk for a particular illness. The case management process can start at that time. *Employee Benefits Plan Review* quotes Laird Miller, former director of Honeywell's health systems department: "In the future, case management will involve more risk identification, assessment and reduction, health promotion, disease prevention, and early intervention."

Employee education can also achieve early intervention. If the case management program has been thoroughly and positively communicated to employees, they will be aware of the option and "refer" themselves to the case manager at the onset of illness. If a company does not perform precertification, the trigger points may come into play at varying times during the life of the claim. *Employee Benefits Plan Review* lists the following guidelines suggested by Henry Orlik, manager of CNA Insurance Company's CURE PLUS program:

- Certain diagnoses that will relate to a medical coding system. The standard coding used in health care today is the ICD-9 (International Classification of Diseases, Ninth Edition).
- Terminal or progressive illnesses.
- A designated claim dollar amount—$5,000 to $10,000, for example.
- Repeated admissions.
- Patterns of outpatient therapies in excess of six weeks.
- Home care by a registered nurse for two to four hours.
- Skilled nursing exceeding six weeks.
- Hospital interim billing. (Often, if the hospital expects a large bill to mount up, it will bill at interims rather than waiting until the patient is discharged.)

All too often, the "alert" does not sound until a patient is approaching his or her lifetime maximum benefit. Both optimal care and optimal cost containment require early intervention.

In the case of AIDS, Gavin Kerr at MONY Financial Services gives the following trigger points:

- More than one admission during a six month period
- *Pneumocystis carinii*

- CNS lymphoma
- Meningitis (cryptoccus)
- Kaposi's sarcoma
- Tuberculosis
- Other opportunistic infections

Identifying AIDS cases is not always simple. Usually, the diagnosis given on medical reports is one of the opportunistic diseases or some other related condition.

Step 2. Planning and Coordinating

The conventional health care delivery system has been called a "patchwork method of delivery and reimbursement." Commonly, catastrophic cases in particular have many physicians involved in a single case. In both the hospital setting and for outpatient treatment, tests are often duplicated and procedures repeated.

Norman Cousins, in his best-selling book *Anatomy of an Illness,* describes an all-too-familiar occurrence in hospitals. He expresses astonishment at being visited in quick succession by four separate technicians from four different departments to draw blood. His solution: "When the technicians came the second day to fill their containers with blood for processing in separate laboratories, I turned them away and had a sign posted on my door saying that I would give just one specimen every three days and that I expected the different departments to draw from one vial for their individual needs."

A unique feature of the case management approach is the coordination provided by the case manager. After assessment, the case manager works with the attending physician, other providers, the benefits manager, community services, and the patient's family and significant others to develop a treatment plan. Communication and teamwork, facilitated by the coordinator, are all-important at this stage.

A good case manager is thoroughly familiar with both community and, if necessary, national resources and with treatment alternatives, although the attending physician always makes the final decision. There are all kinds of treatment facilities that add to the combination of a treatment plan. It is very difficult for any one practitioner to optimize the treatment and minimize the costs. For example, a physician may continue a patient's hospital stay in an acute care facility because he or she is unaware of a specialized rehabilitation resource. A case manager with expertise in many available settings may be able to suggest a more effective alternative.

In the case of AIDS, a case manager might be more familiar with community resources and support structures than the attending physician. In addition, when serious illness strikes, it is often difficult for a patient and his or her family to make rational decisions. A good case manager can offer quality choices and develop an individual, customized treatment plan.

Having discrete elements in place as part of a cost containment program—preadmission certification, a second opinion program, concurrent review, and so on—does not mean a company is providing case management. Case management is distinguished by identifying cases that could benefit from specific intervention and then getting sufficiently involved in the case to assess the needs and develop and coordinate a treatment plan.

Coordination must be timely, say case managers. Administrative delays can jeopardize a treatment plan and increase costs; for this reason, companies should work out quick approval procedures. In this regard, a good case manager builds up relationships with vendors and other providers and can ensure speedy delivery of care.

Step 3: Monitoring Treatment

Whether the patient is in an acute care facility or at home, the case manager monitors his or her course of treatment. Communication continues to be a key element. In the case of AIDS, home care usually provides the type of treatment the patient prefers. In addition, long hospital stays carry risks. Lynne Tonsfeldt, medical case management product manager for Intracorp, told *Employee Benefits Plan Review*: "It is estimated that 1.8 million patients had [in the course of a year] prolonged hospital stays because of infections acquired while in the hospital." For AIDS patients, a depressed immune system makes them highly vulnerable to infection.

Next, a look at what persons interviewed for this briefing indentified as critical success factors.

COMMUNICATION: THE MORTAR THAT HOLDS IT TOGETHER

Effective case management is impossible without good communication. Everything from employees' perceptions of the program to physicians' willingness to cooperate, to the success of the treatment plan hinges on it.

The more employees know about the program and how it works—and the more positive their impressions—the more likely they are to self-refer early in the course of an illness. The program must be visible. Some providers aid in the education effort and rely strongly on employees' word of mouth endorsements to co-workers, family members, and friends. The best providers are offering a real service and are eager to deliver their message.

Case management poses special issues of confidentiality. If the service is provided in-house, employees may have concerns about their employers' having knowledge of their medical history. They must be assured that whether the case manager works in house or via a carrier, the medical information will be held in strictest confidence. If disability or other requirements necessitate communication of a patients' medical condition or diagnosis to an employer or benefits manager, employees must feel sure that any information disclosed will be used for medical or benefits reasons only.

Within the past several years, the medical community has gained greater experience with case management and is learning to view it positively. Case managers can have a role in helping to educate physicians about case management. If a company has a preferred provider organization or if a number of employees use the same doctor locally, the coordinator can provide the physician with written material and answer questions.

Physicians have many reasons to favor case management. The utilization review aspects, however, are more controversial. In administering case management, some providers stress physician support—that is, providing assistance to assure that the best and most efficient resources are secured. Others emphasize monitoring the physician's use of resources. The latter approach may generate some resistance.

THE COORDINATOR: CASE ORCHESTRATOR

The core of the case management approach is the case manager—usually a registered nurse or a physician—who works with the attending physician and the employee to coordinate medical care. The case manager serves as the patient's advocate—with the insurance company, the doctor, other care givers, community organizations, and health care agencies. The goal of this teamwork is to marshal all the resources necessary to offer quality options to the patient and maximize the effectiveness of his or her care. It is always important that the patient and the doctor make the final treatment decisions, say many case workers. (Case managers and insurance providers are not in a position to practice medicine; nor do they claim to be. In the event of a disagreement, a physician team or a physician consultant works closely with the case manager and the attending physician to come to a resolution.)

The more knowledgeable a case manager is, the more creative he or she can be. In one situation, a resourceful case manager arranged motel accommodations for a patient receiving antibiotic infusion, an arrangement that allowed close contact with his family as well as reducing costs considerably. Another persuaded an employer to remove architectural barriers to allow the use of a wheelchair and helped that employee to find new living quarters when his home became unsuitable because of his medical condition. At the same time, he or she routinely arranges for exceptions to benefits by consulting with the employer and the carrier.

Gavin Kerr, director of managed health care at MONY Financial Services, sums up the qualifications needed for this demanding profession: "The best case managers have extensive clinical experience, extensive medical knowledge, exposure to a variety of alternative treatment plans including home care, compassion for the

[Continued on p. 105]

A PROVIDER'S POINT OF VIEW:

MONY Financial Services

The following is an interview in which Anne Skagen, staff reporter, talks with MONY Financial Services' director of managed health care, Gavin Kerr, about case management. MONY is a financial services company, one division of which provides group insurance to employers throughout the country. Much of this business is in San Francisco and New York, areas where the AIDS threat is most significant.

A.S. Do you find that there is as much use of the case management approach on the East Coast as there is on the West Coast?

G.K. There's less sophistication on the part of the policy holder in New York. In terms of companies establishing case management as part of their plans, it's really penetrated both ways. It's pretty extensive. About two-thirds to three-quarters of our businesses have case management programs as part of their plans.

A.S. What have you found physicians' reactions to be?

G.K. Medicine in general is very open to working with case management. It tends to be the utilization review piece that we find physicians balking at. But once the case has been identified as requiring case management, the physicians in many ways really like it because it does two things for them. First, it ties them more closely into the benefits plan so that they know what they have to work with, what will be funded. It provides a neutral third party to work with the insurance company to come up with benefit exceptions that will allow them to do some creative things in home health care or intermediate care or any of the other convalescent care settings. Second, it does provide—and this is something the providers are just catching on to—additional documentation and support in case of a malpractice suit. That's a real plus, especially in a litigious environment.

A.S. Would you comment on in-house programs?

G.K. Some do it. It's tough, especially if you don't have a utilization review program in house, because there is a delay in getting the patient data to whomever's doing the in-house case management. And then there are so many specialized pieces of it so that it's very difficult for a one-person shop to do.

A.S. Is case management vulnerable to liability suits because of the close involvement of the insurer and the employer in decision making?

G.K. Not necessarily, primarily because the ultimate treatment decisions still remain with the physician and the patient. We are a support to that process. We are a resource for that process. We will work with the insurance plan to provide additional resources when appropriate. But the ultimate decision still remains with the physician and the employee. We are very, very careful about this. We are not in a position to practice medicine, nor are we the treating physician. He or she ultimately has the expertise and the knowledge as well as the experience in treating that particular patient.

A.S. Do you get resistance from employers when trying to sell case management?

G.K. Case management is easy to sell. MONY's philosophy is to provide in-house case management in cases where it isn't funded because it's in our best interest. We have our own protocols, but it's not the same program we offer to employers using the utilization management program. But, generally, the hesitancy tends to be with utilization review—will employees accept it, are they willing to do it?

Usually when there are questions along these lines, case management is what sells the program. It just makes such logical sense. Chrysler recently released a statement about case management being the heart of their managed care program. They did a study and found that 3 percent of their claims accounted for 45 percent of their costs. The way they've really controlled their expenses is through their case management program. I think more and more people are finding this.

A.S. Do you have some advice to give employers contemplating a case management program?

G.K. First, I strongly encourage employers to set up a utilization management program—that includes both utilization review and case management, because case management without the up-front flagging is much less effective.

Second, I'd look at the credentials of the organization providing the service. Make sure its people are knowledgeable. Make sure they have expertise in all of the case management areas. Find out what they know about AIDS.

Third, employers have to regard case management as sort of an insurance policy. You're setting up a program that is going to pay off on 3 percent of your claims, so that although you may be paying fees for a while for the service without seeing results, over the course of a couple of years you're going to see big payoffs. For smaller employers, someone with 100 employees, it may take two or three years to get that catastrophic claim, and then there may be three or four in a row, maybe an AIDS case, an accident. So it will pay off. It will pay off very well, but in the first year you may not see it.

A.S. To sum up, what criteria would you use to evaluate the success of a case management program?

G.K. The primary one is quality of care for the patient. We think quality of care is really the most important cost containment piece because if you have people who are being treated effectively, their ultimate claims cost is going to be less. If you have a program that's getting them into the right setting, if you're getting them out of the hospital sooner, then you'll reduce the infection rate and other complications. Any time you have a complication it's tremendously expensive, so that quality care is ultimately the best cost savings.

In addition to that, I'd say that cost savings are obviously a good measure.

The third thing is something that's discounted. That's employee satisfaction. When your employee benefits plan helps somebody in a crisis, that can't help but pay off in terms of your employees' satisfaction with the plan and their loyalty to the organization.

patient, the family, and significant others, the negotiating talent of an oil trader and the wisdom of Solomon."

The best case management programs employ coordinators who are specialists in various medical conditions. A case manager dealing with persons with AIDS, for example, must be an expert in the medical, social, psychological, and financial aspects of the disease and must be familiar with resources available in all these areas.

EFFECTIVE UTILIZATION REVIEW

The catchword here is "effective." A recent issue of *Business Insurance* headlined the following warning: "Few UR Firms Are Effective, Says Consultant." One provider told us, "There are people who think all you need is a bank of telephones and a list." The charges may have merit. A UR company we talked with in Texas used a nurse who had no training as a reviewer and who described her mastery of the job as "seat of the pants." Some firms use sophisticated high tech to assess and monitor their cases. These providers are especially critical of their smaller, less sophisticated competitors. "A small TPA that is offering precertification, a second opinion program, and catastrophic case management and runs the whole shop with a single registered nurse cannot be giving effective service," said one respondent. Obviously, the smaller firms disagree, feeling that they can give more personalized service.

On one point everyone agrees: The range between "good" and "poor" is enormous. Dr. Arnold Millstein, president of National Medical Audits, subjected the records of 200 UR vendors to close scrutiny, using a board of physicians to determine the percentage of hospital days that were medically necessary. The results of the audit, as reported in *Business Insurance,* indicate that the best of the UR firms scored in the 90 percent range; the worst at 51 percent. According to Millstein, the latter score is "equivalent to having no utilization review at all."

Millstein also recommends a look at the UR company's track record: Can it show statistical evidence of having reduced hospitalization for its clients? "The reduction in hospital days per 1,000 covered individuals is by far the most effective measurement," Millstein told *Business Insurance*.

There is, of course, a grey area in utilization review—room for exceptions and subjective judgments. As one vendor told us, "If we give the doctors all that they ask for, we're not doing our job. On the other hand, if the doctors are irate 80 percent of the time, we have to look at what we're doing."

Most consultants and practitioners recommend thorough evaluation of the potential UR vendor, with requests for proposals (RFPs) submitted to a variety of candidates. Central to the evaluation is the scope and degree of medical support within the system. Often overlooked, say some, is the educational support for employees. Will the UR vendor help communicate the program to the workforce in a way that will help employees to understand and accept it?

EXPERT SYSTEMS

Use of expert computer systems in UR represents the cutting edge in identifying and monitoring claims. Medical "rules" or protocols are encoded into the software to guide nurses in making decisions as to hospital admissions, flagging for case management, and the suitability of the patient for the program. In addition, he or she can develop protocols for treatment.

Developing fine-tuned expert medical standards is the crux of achieving results. The question is, as Paul Gertman, M.D., put it, "Are you trying to use a hatchet or a screw driver to get a bolt in and out?" In theory, a hatchet can do it, but it can also do a lot of damage in the process.

The same cautions suggested in contracting a UR organization apply to picking an expert system. According to Gertman, "Selecting an expert system is something like selecting telephones. Some are made by AT&T and Southwestern Bell, and others are made by Taiwanese rip-off shops. They all look alike, with nice plastic on the outside. But the innards are very, very different."

How the System Works

How does a carrier decide whether to provide coverage when a doctor makes a decision related to length of hospitalization, outpatient surgery, and the like? Many carriers now offer services in which, say, all hospitalizations require preadmission review. Some carriers have developed their own "expert systems," computerized information—in a decision tree format—that helps the carrier determine what is appropriate in each case. Such systems are often based on what is known as PAS (Professional Activity Study, conducted by the Commission on Professional and Hospital Activity) data: national averages for length of stay by diagnosis. Often such systems focus more on the administrative than on the clinical, although some medical knowledge may be part of the program.

The most extensive such system to date is known as Optimed, developed by The Health Data Institute of Cambridge, Massachusetts. Optimed is available directly from HDI, or a carrier may license the software.

The step-by-step procedure for using the service is somewhat the same, regardless. A doctor tells a patient that he will require hospitalization; the patient tells the doctor that his carrier requires a utilization review. The doctor or his nurse then calls the carrier or Optimed, where a service person checks the patient's policy provisions and then enters questions into the computer related to the diagnosis and the patient's general condition and indications.

If the coverage specifies the need for a second opinion, the system guides the representative to ask for indications that may overrule that provision. Finally, the system will ask the physician the length of stay requested and check this information against recommendations built into the system. Again, there is opportunity to indicate conditions that would call for exceptions.

The carrier, of course, pays for the system—either by developing its own, by leasing software, or by making provisions for an organization such as HDI to do the review. And those costs are passed on to the company. Over time, however, monies saved will more than cover the cost of developing and/or using the system.

HDI utilized a variety of resources in developing Optimed. A clinical database of 8 million cases provided a starting point. The organization worked with small groups of doctors—those recognized as specialists in different branches of medicine—to formulate the basic questions and protocols for the system; this input was checked against the PAS averages and the information in the clinical database, then reviewed by a panel of 150 physicians nationwide.

Buyer Beware

What questions should a buyer ask of a vendor who is offering UR services in general and an expert system in particular? We put this question to Paul Gertman, M.D., formerly of The Health Data Institute. His recommendations:

1. What is the medical expertise behind the system? How many doctors were involved in creating it? How much time was spent on it? What's the level of protocols and medical "rules" in the system? How many are there? Is there one screen? Or are there 500? Anyone who tells you that you can have one simple set of rules for doing real-time utilization control can't be trusted.

Who was involved in the design? What are their credentials? And what is their track record? You should ask whether there's a team of nationally recognized experts behind it.

2. How do they deal with exceptions to the protocols generated by the system? Do they have an appeals process? Who stands behind it? Who reviews it? If there's a dispute with the physician, what recourse do patients and their physicians have? Is there a medical review process, an appeals process by independent experts?

3. Can the company provide you with results? Don't rely only on what it claims. Can it provide you with their claims experience? Can it document?

4. What's the process for updating the system? Does the company have one? It has to be able to demonstrate that it can go back in and refine and enhance the system. The world is changing. Treatment is changing. Constant refinements are essential.

5. What is the caliber of the training and review of personnel? What are the company's internal quality controls? This issue is

highly important. A good quality control process can mean the difference between success and failure.

If you can come up with positive answers on all these points, you're probably buying an effective program.

LIABILITY

None of the carriers or third parties we spoke with thought there were any special liability issues connected with case management or utilization review. Providers generally carry insurance to cover potential lawsuits. But all stressed that they were very careful in their decision making. Aetna, for example, draws up an agreement to which all parties involved in a case must be signatories. The company allows the agreement to be severed at any point.

There are also legal issues centering on confidentiality and right to privacy. Dr. Brian Latham, in *Health Care Costs: There Are Solutions* (AMACOM), succinctly sums up the issue of an employer's right to medical data:

"Insurance laws vary from state to state, but in most U.S. states employers have a legal right to obtain health care utilization data on their employees. If the company is self-insured, there is no problem. If it is not, it is likely that the company will have to gain information from its insurers, agreeing that it will be used solely to investigate health care utilization in the community and not to penalize employees. In short:

- "There are no legal restraints preventing an employer from collecting or securing medical cost and utilization data in an effort to contain costs.
- "However, employers are legally prohibited from misusing the information to penalize employees who are high-cost users of medical care. . . "

To this we might add, "or for any other reason. The data cannot be used for purposes that might be construed as discriminatory." ERISA guidelines, state insurance codes, laws governing health care, employee benefits, utilization review, and record retention require compliance, and an employer interested in adopting case management or utilization review should consult legal counsel.

Here are some general guidelines suggested by industry spokespersons:

- Make the program voluntary.
- Offer it to everyone, not just the management group.
- Give the patient choices.
- Don't endorse certain facilities or certain procedures.
- Be sure that a physician is making the decisions, not a lay person.
- Build flexibility into the design of the benefits package.

A California appeals court case, *Wickline* v. *State of California*, involved the first major malpractice suit against a third-party payor. Although the decision favored the review organization, the appellate opinion stressed that "third-party payors can be held liable for cost containment decisions that deprive patients of medical care."

The decision, while it occurred in California, will undoubtedly influence future malpractice claims that could apply to both employers and utilization review firms.

PAYBACKS

Catastrophic cases have traditionally been poorly managed. The case manager, by adding his or her expertise, can introduce efficiency, often accelerating care and allowing an employee to return to work more quickly. The same kind of efficiency tends to mean fewer complications, such as secondary infections. In addition, the coordinating the case manager provides prevents duplication of services.

The cost containment aspects of case management can be quantified. The largest savings come through offering alternative treatment settings. In many cases, the most appropriate care may also be less expensive. In the case of home care for persons with AIDS, the savings are three to fivefold and sometimes greater. On the other hand, short-term expense may be greater, with long-term savings. This is particularly true for rehabilitation facilities specializing in a particular condition. Although more expensive than another setting in the short run, the expertise available may hasten a person's recovery allowing a speedier return to work. The most commonly cited ratios of savings to cost range from 4 to 1 to 10 to 1; that is, a savings of from $4 to $10 on every $1 invested.

COSTS OF INSURANCE

Johnson & Higgins, benefit consultants, gives an interesting and highly useful prediction model showing how an employer can determine the number of AIDS cases likely to occur in its employee group.

In underwriting, insurers rely on the demographics of the group and the actuarial value of the risk associated with those demographics.

Relevant demographic data for AIDS include:

- Sex
- Age
- Geographic location
- Sexual preference
- Drug use

According to CDC estimates, between 1 and 1.5 million Americans carry the AIDS antibody (would test HIV positive). For the population 20 years old and above, this is roughly 8 in every 1,000 persons. The employed population is estimated at ages 20 to 59. Within this group, the HIV infection rate for men is 18 per 1,000 and for women, less than 1 per 1,000.

On the basis of data collected so far (and this is just a preliminary estimate), it is thought that the prevalence of HIV infection among insured individuals (both group insurance and individual) is about half that of the general population. Therefore, for insured men, 9 out of 1,000 would test positive and for insured women, 0.5 out of 1,000. The most recent data on the conversion rate (those who test positive and eventually develop full-blown AIDS) is 36 percent. (Sadly, it is now believed that all of those infected may eventually progress to full-blown AIDS). Johnson & Higgins uses the 36 percent figure in this calculation.

Geography influences the probable number of AIDS cases an employer might expect. Five states are high-risk areas: New York, California, Florida, Texas, and New Jersey; and two cities, New York and San Francisco. Over 70 percent of all cases have occurred in these four states and 35 percent in those two cities.

National statistics on homosexuality indicate that 3 percent of adult males are gay and another 3 percent bisexual (during some part of their lives). Out of these two groups combined, 30 percent are estimated to be HIV positive.

It is believed that the vast majority of intravenous drug users are not employed or insured.

Exhibit 1 takes a hypothetical company, Sports Distributors, Inc.,

Exhibit 1. Estimated AIDS risk for Sports Distributors, Inc.

Demographics: 10,000 employees in 35 states; ages 20 to 59

7,000 male	5,500 married
3,000 female	1,800 married

Male participants		Female participants		Total
Employee	7,000	Employee	3,000	
Spouse	1,800	Spouse	5,500	
Total	8,800	Total	8,500	17,300
HIV infected rate	@ 9/1,000		@ 0.5/1,000	
HIV positive	79		4	83
AIDS rate	36 percent		36 percent	
(7 years) AIDS cases	28		1	29

with an employee population spread nationally, and analyzes its AIDS risk. The significant demographics here are age and sex.

Local Financial Services, Inc. is a service provider located in a major metropolitan area. Here we are concentrating on the male employee participant. (See Exhibit 2.)

Exhibit 2. Estimated AIDS risk for Local Financial Services, Inc.

Demographics: 10,000 employees in metropolitan areas; ages 20 to 59

7,000 male		5,500 married		
3,000 female		1,800 married		
Male participants		Female participants		Total
Employee	7,000	Employee	3,000	
Spouse	1,800	Spouse	5,500	
Total	8,800	Total	8.500	17,300
Male homosexual @ 3 percent				319
Male bisexual @ 3 percent				519
				1,038
HIV infected		30 percent		311
AIDS rate — 7 years		36 percent		
AIDS cases			112	

6

The Legal Issues: What Every Manager Should Know

By Arthur S. Leonard

The AIDS epidemic raises a myriad of significant legal issues for employers, ranging from hiring, promotion, assignment, and transfer practices to employee benefits (including leave policies), the impact of customer and co-worker reactions on personnel, decision making procedures for dealing with medical information about employees, and the circumstances for terminating employment. In this chapter, I will discuss the central legal principles relevant to these issues, and will attempt to provide guidance on applying those principles to the particular problems that companies have to face on a day-to-day basis.

MEDICAL ASSUMPTIONS UNDERLYING THE LEGAL ANALYSIS

The legal issues raised by AIDS cannot be addressed in a vacuum. Certain assumptions about the medical facts have been made throughout this chapter, and are summarized briefly here.

Acquired immune deficiency syndrome (AIDS) is the result of complications of infection by the human immunodeficiency virus (HIV), a blood-borne virus that is primarily attracted to certain white blood cells called T-4 helper cells, although it has also been found to infect some other human cells, including cells of the central nervous system. The virus cannot survive long outside a liquid environment, and proven transmission mechanisms thus far include only blood and semen, although small amounts of the virus have been found in saliva, tears, and vaginal secretions. Studies of family members of infected persons, as well as of large numbers of health care workers who have cared for persons with AIDS, have provided convincing evidence that casual contact (and even relatively close and continuing contact) will not result in transmission of the virus.

Within a few weeks to six months of infection, an individual may develop mild flu-like symptoms, which are characteristic of the initial response when the body forms antibodies to the virus. After this initial response, infected persons may suffer no ill effects for many years. Some, however, may develop such physical symptoms as fatigue, swollen lymph glands, fevers, or weight loss, which, collectively, have been designated AIDS-related complex (ARC). Others fall prey to opportunistic infections and rare cancers such as Kaposi's sarcoma, and thus meet the official definition of AIDS.

Many persons with ARC or AIDS have significant periods of relatively good health when they are able to work. Not all those diagnosed as having ARC proceed to AIDS, although some people have actually died from the symptoms associated with ARC without ever meeting the official definition of AIDS. Persons with AIDS live for varying periods of time, and some people so diagnosed as long as five years ago are still alive today and physically able to work. However, most people diagnosed as infected with AIDS have died from complications of the syndrome within two years of diagnosis, and many of those have died within 18 months. The main cause of death for most of these individuals has been *Pneumocystis carinii* pneumonia. The opportunistic infections associated with AIDS are caused by infectious agents that normally do not cause illness in human hosts. This is why it is highly unlikely that a healthy person will catch *Pneumocystis* pneumonia or other AIDS-related infections from a person with AIDS. In fact, most adult Americans who live in urban areas are chronically infected by many of these infectious agents, yet never suffer any ill effects from them.

Consequently, the key medical facts in considering HIV infection and the workplace are:

1. HIV is not casually transmissible in the workplace, and opportunistic infections associated with AIDS do not present an appreciable risk to healthy co-workers of persons with AIDS.
2. Most persons infected with HIV will not become ill within the first five years of infection, and many may remain healthy for much longer periods.
3. Many persons with AIDS will enjoy substantial periods of relatively good health during which they are able and willing to work.

THE CENTRAL LEGAL CONCEPT: AIDS AS A HANDICAP

The central legal concept in thinking about AIDS issues in the workplace is the idea of AIDS (and any potentially disabling disease) as a "handicap" or "disability" within the meaning of pertinent federal, state, and local laws. Beginning in the 1970s, the federal government and almost all the states (as well as some large municipalities) have enacted laws affecting the ways in which employers may deal with employees who are (or are perceived to be) handicapped or disabled. These laws generally define "handicapped individuals" quite broadly as people who have a physical or mental impairment that may significantly affect their ability to function on a daily basis (including their ability to hold a job), and require that personnel decisions with respect to such individuals be made on a rational, nondiscriminatory basis. Thus, a person who suffers from a handicapping condition but is physically and mentally able to perform a job without significant risk of harm to himself or others may not be the object of invidious discriminatory treatment. In certain kinds of jobs, employers may even have a statutory or contractual obligation affirmatively to seek out and employ qualified handicapped individuals.

It is possible that the handicap discrimination laws will be superseded with respect to AIDS and HIV infection by pending federal legislation during 1988, but the legislative proposals to specifically ban such discrimination face opposition from the White House, making final enactment questionable.

TO WHOM DO HANDICAP LAWS APPLY?

The federal handicap discrimination law, known as the Rehabilitation Act of 1973, imposes different obligations depending on the following classifications:

Federal agencies and programs: Section 501 of the Act requires federal agencies and programs to adopt affirmative action programs to seek out and employ qualified handicapped individuals. Section 504 requires that no federal agency or program deny otherwise qualified handicapped individuals the benefits of participation in the agency or program, including employment.

Federal contractors: Under Section 503, contracts in excess of $2,500 between private employers and the federal government must include affirmative action requirements with respect to qualified handicapped individuals.

Federal funding recipients: Any enterprise, whether public or private, receiving direct federal financial assistance must refrain from excluding or discriminating against otherwise qualified handicapped individuals, by virtue of Section 504.

Because state laws vary in their applicability, the laws of the jurisdictions where you do business should be consulted directly. As a general rule, most state laws apply to all public and private employment in the state, with occasional exceptions specified in the laws. In a small minority of states, laws against handicap-based employment discrimination apply only to public sector employers. All the larger industrial states, which are (as of 1988) the leading sites of the AIDS epidemic, require private employers not to discriminate unfairly against the handicapped.

City ordinances apply only within the geographical limits of the enacting municipality, but present an additional consideration: some cities have enacted AIDS-specific laws, and these should also be consulted together with the more general state and federal laws. Most of these cities are in California. In addition, many large cities, as well as the state of Wisconsin, forbid employment discrimination on the basis of sexual orientation, a frequent adjunct to AIDS-based discrimination.

WHAT IS REQUIRED FOR EMPLOYER COMPLIANCE?

The first step for compliance with a law requiring *affirmative action* is, of course, to adopt an affirmative action plan. Presumably, those employers who have such contractual obligations under federal (or some state or city) contracts will already have in place affirmative action plans that include handicapped individuals. Employers should review these plans to determine whether modifications are needed to include persons affected by AIDS. In many large cities, there are AIDS assistance organizations that would welcome the opportunity to refer qualified persons with AIDS or ARC or persons who have tested seropositive and who are in need of part-time or full-time work. Employers under affirmative action obligations would be performing a distinct public service (and incurring significant gratitude and loyalty) if they were to identify suitable jobs and employ such persons.

Personnel Procedures Affecting Hiring, Assignment, and Promotion

For most employers affected by handicap discrimination laws, there may be no express affirmative action requirement, merely a nondiscrimination requirement. Personnel policies and practices should be reviewed to ensure that they do not discriminate inappropriately against persons affected by AIDS.

For instance, in hiring procedures, inquiries of applicants about their physical and medical condition should be designed to elicit only information relevant to prospective job duties. HIV, the viral agent believed by most researchers to be a necessary element in causing AIDS and AIDS-related complex, is not casually transmissible, and it appears that only a minority of infected persons will actually become ill as a result of HIV infection; therefore, the relevance of HIV infection to *present* ability to work is slight. Applicants with ARC or AIDS may be suffering from symptoms (including fatigue, weakness, or specific opportunistic infections) that are relevant to some aspects of job performance; thus it may be appropriate to inquire about those physical impairments to the extent that they relate to actual job requirements.

In assignment, transfer, and promotion procedures, the present ability of the employee to perform the job in question is the key factor under handicap discrimination law. Speculation that an employee may in the future become unable to perform the job is generally not considered relevant, unless expert medical opinion indicates that performing the job will present an *immediate* danger to the health or safety of the employee or to those around him or her. Consequently, inquiries or medical examinations that may accompany decisions about assignments, transfers, or promotions should be tailored to the acquisition of information relevant to present ability to work or immediate dangers to health and safety.

As with hiring decisions, the mere presence of HIV infection, unaccompanied by any physical symptoms of disease, would not be relevant to most assignment, transfer, or promotion decisions. If an employee is suffering from ARC or AIDS, an individualized assessment of his or her physical abilities and medical prognosis would be relevant. For example, promotion to a high-stress position or a position calling for sustained physical exertion could present a health challenge that required the advice of a physician knowledgeable about AIDS.

But certain fears commonly expressed by employers do not seem relevant once the facts about AIDS are understood. For example, in one case an employer wished to exclude an employee with ARC from an office setting, arguing that inevitable paper cuts suffered by clerical workers could result in virus transmission. There is no medical basis for such an argument.

Reasonable Accommodation

Many of the applicable laws require employers to make a "reasonable accommodation" to the handicaps of otherwise qualified employees. Sometimes this can be done simply by installing ramps to make a workstation wheelchair accessible, or by adopting a flexible approach to attendance rules for employees whose travel to work is easily impeded by bad weather conditions due to the physical limitations on their mobility.

In other cases, accommodation could be costly and require major restructuring of a workplace. The degree of accommodation required by the law is a judgment call in any particular situation; there are few hard-and-fast rules. But the Supreme Court has indicated a few principles, and others are suggested by administrative regulations.

1. The offer to transfer a handicapped employee to another job may be made by an employer as an accommodation, but it is not legally required unless the employer has a regular policy of offering job transfers to meet the needs of individual employees.
2. The degree of expense and inconvenience that may be required to accommodate a handicapped person is proportional to the size and scope of an employer's operations. A large employer with thousands of employees and job classifications can be expected to absorb more in the way of costs to accommodate an individual employee than can a small employer with only a handful of job classifications.
3. In any event, a major expenditure disproportionate to the employer's overall operating budget is not required.

In thinking about the concept of accommodation, perhaps the major emphasis should be on the word "reasonable." We are dealing with situations calling for flexibility, imagination, and a willingness to innovate in order to meet human needs. The employee requiring an accommodation also has a responsibility to be flexible and to appreciate the employer's business needs in arriving at a workable solution to the problem. If the employer discusses the need for accommodation in a calm, matter-of-fact manner with the employee, it is likely that an arrangement can be worked out that is beneficial both for the business and the employee.

One constantly recurring issue with respect to AIDS is whether a transfer to a nonpublic contact position can be required as a condition of continued employment. If an employee is physically able to work in his or her normal public contact position, the law would technically require that the employer retain the employee in that position. However, if on the basis of actual experience it appears that there is a significant negative impact on the business, a transfer after a frank, calm discussion with the employee may be an appropriate solution to the problem. Persons without physical symptoms should be evaluated on an individual basis, and such decisions should not be based on mere speculation about possible customer reaction.

Employers have a strong financial incentive to evaluate their leave policies in light of the AIDS epidemic. In some circumstances, paid medical leaves will be a welcome alternative to an employee suffering from ARC or AIDS, and will prevent workplace disruption. It will also reinforce employee morale to know that such leaves are available in case of medical emergencies. However, an employee who is able to work and who wants to work may not be forced to take a medical leave unless the employer can prove that removal of the employee is absolutely necessary for the safe operation of the workplace.

AIDS AND HANDICAP LAW: THE REACTION OF THE COURTS

Writing in 1985, when the AIDS problem in the workplace was relatively new, I predicted that the courts would apply principles of handicap discrimination law in the ways described above. Developments since then have tended to confirm my prediction. In decisions involving schoolchildren infected with HIV who are seeking admission to the classroom, the entitlement of persons with AIDS to equal access to public accommodations, and some cases of AIDS-related employment discrimination, the courts have uniformly treated AIDS as a handicap under applicable employment discrimination laws, and have focused their inquiry on whether the individual in the case was otherwise qualified for what was being sought.

Perhaps the most important court development since 1985 has been the Supreme Court's decision in *School Board of Nassau County, Florida* v. *Arline* (1987). In this case the court held that persons impaired by contagious diseases should be considered "handicapped" under the federal Rehabilitation Act, and thus entitled to protection from discrimination if they were otherwise qualified. The case concerned an elementary schoolteacher who had suffered from recurrent tuberculosis. The court agreed with a ruling by the Eleventh Circuit Court of Appeals that the trial judge should evaluate expert testimony to determine whether there was a sound basis for excluding the teacher from the classroom, and rejected arguments advanced by the U.S. Justice Department that personnel decisions motivated by fear of infection could not be classified as "handicap discrimination."

Because the schoolteacher in *Arline* had actually been impaired by tuberculosis, the Court had no need to express a view as to whether infected individuals who are not physically impaired should be considered handicapped. However, several other courts have answered that question affirmatively, most notably the United States District Court for the District of Columbia in *Local 1812, American Federation of Government Employees* v. *U.S. Department of State* (1987). In this case Judge Gerhard Gessell ruled that State Department employees who were infected with HIV should be considered handicapped within the meaning of the statute. (He also concluded that the special needs of the State Department rendered such individuals not "otherwise qualified" for overseas diplomatic assignments.)

The most important legal developments at the state and local levels have included a Massachusetts court decision in favor of a telephone lineman with AIDS who had been constructively discharged due to panic by co-workers, and a ruling by the California Fair Employment and Housing Commission which suggested that enforcement officials should seek temporary injunctive relief to get persons with AIDS returned to work pending the outcome of investigations of their discrimination complaints. Numerous state agencies have announced their position that persons with AIDS and ARC (and, in most instances, symptomless persons infected by HIV) are protected from discrimination. In Wisconsin, a state agency ruled that a policy of excluding people with AIDS violated provisions of state law forbidding discrimination on the basis of sexual orientation, because of the adverse impact such a policy would have on gay male employees.

While federal courts have rejected challenges to the HIV testing and exclusion policies practiced by armed forces and the Foreign Service of the State Department, they have upheld challenges to AIDS-based discrimination against other public employees and schoolchildren. It seems likely that the special circumstances pertaining to the military services (which are not covered by the Rehabilitation Act) and to the Foreign Service are responsible for these decisions, not any general predisposition of the federal courts against extending protection to persons with AIDS under federal handicap discrimination law. Private sector employers therefore should not rely on the military and Foreign Service cases as precedents to justify discrimination.

AIDS AND EMPLOYMENT LAW: PRACTICAL CONSIDERATIONS

Apart from handicap discrimination law, other principles of employment law are applicable to the AIDS situation, and should be borne in mind by the careful manager. Some of the most important are discussed below.

Wrongful Discharge and Other Common Law Actions

One of the most significant new areas of employment law is the rapidly developing litigation over wrongful discharge. Lawsuits, generally in state courts, are brought by a discharged employee asserting that an employment contract was violated or that the discharge constituted a tort making the employer liable to punitive damages. Although some state courts (most notably those of New York) have been hostile to such lawsuits, a majority of the states now recognize certain legal theories governing wrongful discharge. The most widely accepted theories have to do with violations of job security promises or of disciplinary procedures set forth in personnel manuals, and with discharges that undermine a public policy set forth in state or federal law. In several states, most notably California, such lawsuits are tried before juries and have resulted in rather large punitive damage awards.

The manner of discharge may itself lead to charges of invasion of privacy, defamation, or libel. State courts seem increasingly receptive to the idea that employees have certain privacy rights against their employers, which may be breached by subjecting them to intrusive tests (such as the polygraph or urinalysis for drug use) or disclosing confidential information about them. Responding to these developments, some employers have concluded that inquiries about employees from third parties require a cautious response, normally limited to confirming that the employee works for the employer and revealing little else. Because of the panic surrounding AIDS, the information that an employee is infected with HIV or actually suffering from AIDS should be treated as confidential, because revealing such information without the employee's authorization could result in common law liability (or, in some states, liability under statutes governing personal privacy or the confidentiality of medical records). A Massachusetts court has held that a company may have violated Massachusetts privacy laws when a supervisor revealed information about an employee's possible ARC diagnosis which had been given to him in confidence.

Consequently, any discharge or personnel action involving an employee with HIV infection or AIDS should be approached with great caution, and alternatives considered, if the prudent manager wants to avoid significant litigation expense and bad publicity. Sustainable discharges must be for documented misbehavior or poor performance, and employers must be consistent in applying reasonable standards of behavior and performance, if they hope to emerge unscathed from wrongful discharge litigation in jurisdictions such as

California or New Jersey. Transfers, suspensions, forced leaves, and similar steps should also be approached with caution, given the legal requirements of accommodation imposed by handicap discrimination law.

Managers should also be aware that during 1987 the state of Montana became the first in the nation to adopt a wrongful discharge statute, extending to all employees in the state the right not to be discharged except for legitimate, work-related reasons. The Montana statute, similar in many respects to the recommendations of organized bar associations and legal academics, may be a harbinger of future legal developments. The prudent manager would be wise to adopt rational personnel decision making and documentation procedures in advance of legislative developments in other states. At a minimum, these would include published work rules, a system of progressive discipline, and careful instruction of line supervisors on the need for consistent enforcement of management policies.

Thus far, discharges on account of AIDS have resulted in a significant number of wrongful discharge filings in California, with many sizable settlements. No such case has yet advanced to the stage of a reported, published judicial opinion.

Employee Benefits Law

The Employee Retirement Income Security Act (ERISA) may have a significant impact on AIDS-related cases in at least two respects: (1) ERISA has its own employment discrimination provision, Section 510, which may be invoked by the U.S. Department of Labor or by individual employees under certain circumstances; (2) 1986 amendments to ERISA extend to discharged employees various rights with respect to continued participation in employment-related health insurance plans.

ERISA Discrimination Law: Section 510 forbids discriminating against employees because they claim benefits to which they are entitled under employee benefit plans. It also forbids discharging an employee in order to prevent the employee from attaining benefit rights. Both these prohibitions may be relevant to AIDS-related discharges. As interpreted by the federal courts, a discharge occurring shortly after the employer learns that an employee has applied for insurance benefits for a potentially expensive disease can give rise to an inference of unlawful discrimination. While the main purpose of these provisions seems to be prevention of discharges on the eve of vesting dates under pension plans, employees have invoked them successfully under other circumstances as well. For example, a discharged employee suffering from degenerative arthritis was found to have a right to sue under Section 510 by the United States Court of Appeals for the Third Circuit in *Zipf* v. *American Telephone and Telegraph* (1985).

Continuation Coverage: Sections 1161 et seq. of ERISA require that most employers who provide group health plans for their employees make available a continuation coverage option for employees who are discharged (except for those discharged due to gross misconduct). The employer may require the former employee to pay premiums for the continued coverage, but such premiums may be no more than 102 percent of the costs of providing group coverage to employees. Of course, as part of the group, the employee's covered expenses will be charged against the employer's experience rating, if health coverage is purchased from a private insurance company.

Occupational Safety and Health Act (OSHA)

OSHA imposes a general duty on employers to provide safe and healthful workplaces for their employees, and a specific duty to observe health and safety regulations.

Because HIV is not casually transmissable in the workplace, the presence of an employee infected by HIV would not seem to violate the general duty requirement. However, in work situations where it is likely that employees may be exposed to the blood of infected persons, special precautions may be indicated. During the spring of 1987, the Centers for Disease Control announced that in a few instances health care workers had become infected as the result of occupational exposure to the blood of infected patients. This announcement led to calls for an OSHA regulation governing exposure to HIV in health care institutions. In August 1987, the CDC issued new Guidelines for the Prevention of HIV Transmission in Health Care Institutions, which the Occupational Safety and Health Administration announced would be incorporated in special health and safety regulations governing such employment settings.

In essence, the CDC Guidelines and anticipated OSHA regulations will incorporate well-known procedures for preventing the transmission of blood-borne pathogens such as Hepatitis B Virus (HBV), and make observance of such procedures a legal duty for all health care employers and their employees. This responds to the complaints of some health care workers and their unions that they have received inadequate instruction in infection control procedures, and that necessary supplies for effective infection control are sometimes not made available by health care employers.

Laws Governing Union-Management Relations

The National Labor Relations Act, as amended, also relates to the rights of employees in the AIDS situation.

If employees are represented by a labor organization, the employer may be required to negotiate with the union about many AIDS-related issues, including the adoption of testing requirements or the ability to offer transfers out of seniority order so as to accommodate handicapped employees. If the employer and the union are parties to a collective bargaining agreement, the agreement may dictate the employer's responses to a variety of questions posed by AIDS, including the circumstances under which medical discharges will be granted, assignments made, benefits collected, and the like. Thus, employers who deal with union representatives would be wise to initiate discussions with the union so as to establish a working relationship that is capable of handling AIDS issues when they do arise.

Employees who are not well informed about the ways in which HIV is or is not transmitted may express great reluctance to work with a fellow employee who is known to be infected with HIV or to have AIDS. Employees have also been known to object to carrying out work assignments where they perceive that customers or clients of the employer may be infected. Federal and state labor laws that protect employees from being disciplined when they engage in concerted activities may affect the ability of the employer to counter such refusals. Section 8(a)(1) of the federal law has been interpreted to mean that employees are protected from discipline when they collectively refuse to work due to concerns over health or safety in the workplace, provided they are acting in good faith. The right to engage in such protected activity may be waived, however, when a union representing the employees agrees to include a grievance arbitration procedure in its collective bargaining agreement, although Section 502 of the Labor Management Relations Act may override such a waiver if the union has objective evidence that a serious danger to employee health or safety exists. Individual employee refusals to work may also be protected under Section 502, although there is scant case law on the question. Regulations issued by the Occupational Safety and Health Administration may also protect employee work refusals, but impose an objective standard that employees will have difficulty meeting in most workplaces due to the limited ways in which HIV is transmitted.

Although the National Labor Relations Board has yet to issue any rulings with respect to AIDS issues, several labor arbitrators have ruled in cases where AIDS was an issue under collective bargaining agreement grievance procedures. Thus far, arbitrators dealing with employee discharge situations have refused to bow to panic about AIDS. Two cases of discharges of employees with AIDS have resulted in reinstatement orders. A case involving a prison guard discharged for refusing to conduct pat searches of prisoners without gloves due to fear of AIDS resulted in a reinstatement without back pay where the arbitrator found prison authorities to be at fault for not providing appropriate education for the guards. In another case, applying a contract provision requiring a prison to notify guards of the presence of prisoners infected with contagious diseases, an arbitrator ordered HIV-antibody testing of prisoners; the arbitrator's order has been upheld, at least initially, in a court challenge.

HIV TESTING AND THE WORKPLACE

Since 1985, various tests have become available to determine whether individuals are infected by HIV. The tests most commonly used are actually not direct tests for the presence of the virus, but rather tests to determine the presence of antibodies, substances produced by the body in response to infection by a foreign agent. Researchers have determined that when a person is infected by HIV, that person will produce antibodies within a few weeks to six months, although in some cases it may take as long as a year after infection for antibodies to be detected. Many common laboratory tests for sexually transmitted diseases are actually antibody tests, so the so-called "AIDS tests" are no exception to the general rule in this regard.

The ELISA test, the most commonly used and least expensive test, is very sensitive to the presence of HIV antibodies, which means that it will almost always produce a positive result if the antibodies are present. However, considering that many other conditions can also produce positive test results in circumstances where there are no HIV antibodies, and that persons tested shortly after infection may produce negative results because antibodies have not yet formed, the ELISA test should not be relied upon in the absence of confirmation and retesting. One authority on the medical screening of workers, Professor Mark Rothstein of the University of Houston School of Law, has written that in a general population (i.e., a population not made up primarily of those at high risk for AIDS due to their sexual or drug-using behaviors), many false positive test results will be produced for each true positive result.

In the most careful testing programs, such as that used by the military, the ELISA test is normally repeated and then confirmed by a more complicated and expensive process called "Western blot" before the conclusion is drawn that the individual actually has HIV antibodies in his or her blood. In studies where the complicated process of "culturing" the virus from blood has been used to evaluate the accuracy of the HIV antibody tests as an indication of actual infection, scientists have determined that the tests are highly predictive for the presence of actual live virus, even though live virus cannot be produced from every blood sample that has tested positive for antibodies.

Still under development but soon to be licensed will be antigen tests for HIV, in which the test chemicals will react directly with proteins on the coat of the virus to give a more definite indication whether the virus is present in the test subject. There is some speculation that the antigen tests may also make it possible for doctors to determine which infected persons are most likely to develop AIDS. The meaning of a positive test (whether an antibody test or an antigen test) for the future health of an individual is still speculative at this time. Because AIDS has been recognized as a distinct medical condition only since 1981, and retrospective studies of stored blood samples show that HIV has probably been present in the American population only since the late 1970s, it is difficult to project what proportion of infected individuals will go on to develop ARC or AIDS in the long run on the basis of current data. Insurance actuaries have been able to calculate, on the basis of existing studies and experience, that a person infected with HIV presents a *short-term* mortality risk too expensive for the profitable sale of life insurance at reasonable rates, and some studies of groups of individuals over periods of five years indicate that some proportion less than half are likely to develop medical complications over that time period, but these predictions must be related to the purposes for which they are being made in order to determine whether the test is an appropriate tool for life insurance underwriting or medical diagnosis and treatment.

For workplace purposes, the HIV antibody tests now available do not produce information that is usable for personnel decision making. Individuals who are infected with HIV would most likely be considered "handicapped" or "disabled" under the employment discrimination laws discussed above, and thus be protected against adverse personnel actions unless their medical condition warranted removal from the workplace or exclusion from a particular job. Regulations issued under the federal Rehabilitation Act restrict the use of medical tests to procedures that uncover disqualifying conditions, and state handicap discrimination laws have received similar interpretations. In addition, a few states have adopted laws banning the use of HIV antibody tests in employment; prominent among them are California, Florida, Massachusetts, and Maine.

In *Local 1812, American Federation of Government Employees* v. *U.S. Department of State,* a union representing certain State Department employees of the Voice of America challenged the legality of an

HIV antibody testing program adopted by the Secretary of State. The union charged that the testing program violated the constitutional rights of privacy and equal protection of its members, who are public employees, and that it also violated Sections 501 and 504 of the Rehabilitation Act. In his opinion dismissing the suit, Federal District Judge Gerhard Gesell ruled that the tests could be used by the State Department because seropositivity (a positive test result) would actually disqualify an individual from overseas assignment for a variety of reasons peculiar to the foreign service, so such individuals could not be considered "otherwise qualified." For most public and private sector jobs, in which seropositivity is not a disqualifying factor, however, such a testing program could violate the Rehabilitation Act or applicable state and local laws.

Apart from their legality, the HIV antibody tests present significant dangers to employers because they generate sensitive information whose revelation may subject the employer to substantial liability. Any employer contemplating an HIV testing program should carefully consider whether the information it produces is worth the potential liabilities, and the care that would have to be taken to produce accurate test results and then to keep such results confidential.

FOR MORE INFORMATION

This survey of the legal issues created by AIDS in the workplace is necessarily brief. Publications of varying quality on the subject have proliferated over the past two years. Perhaps the most extensive in-depth recent consideration of AIDS legal issues (including employment) is contained in *AIDS and the Law: A Guide for the Public,* edited by Harlon Dalton and Scott Burris and published by Yale University Press, 1987, which has been written so as to make the legal issues comprehensible to concerned non-lawyers in business and government.

APPENDIX A: Workplace Testing for HIV Infection

A November 1987 American Management Association survey of 995 human resources managers found that 61 respondents, or 6.1 percent of the sample, offered a test for HIV infection to new or current employees.

Most of this testing is voluntary, offered to employees who wished to be checked for the AIDS virus. But 12 respondents have made HIV-antibody testing mandatory. Nine of them test all newly hired personnel, and two more test some new hires on the basis of job category.

Five of the organizations that require testing told us that applicants who test positive for HIV antibodies are not hired. One respondent—a manufacturer and national defense contractor—said that they terminate any current employee who tests positive.

Thus, of our total respondent base, 1.1 percent use HIV-antibody testing in the hiring process; of all respondents, 0.5 percent would make a job offer contingent on negative results. (See Chapter 6 for the legal implications of this practice.)

Two respondent firms have compulsory testing for all current employees, and two more test selected personnel, one according to job category, and the other on a "for cause" basis—that is, a "reasonable belief" that the employee may have come in contact with HIV.

The question was posed as follows: "Do you offer a test for AIDS [HIV] infection?":

	Voluntary	*Compulsory*
To all current employees	29	2
To selected current employees	22	2
To all new hires	11	9
To selected new hires	5	2
On promotion—all employees	7	0
On promotion—selected employees	0	1

Twenty-nine of the 61 firms that answered positively in any category are health care providers—hospitals, clinics, laboratories, and other health service organizations. But none of these health care providers had mandatory testing. Rather, of the 12 respondents that required tests, 7 were manufacturers, 3 were military units, and 2 were service providers.

The military units reported that "HIV positives" are counseled, given restricted assignments, and referred to a medical center for treatment as deemed appropriate.

None of the respondents offering voluntary testing refused employment to applicants who test positive for the HIV antibodies, nor do they terminate current employees with AIDS. Forty percent refer current employees for treatment, and 6 respondents reassign "test positives" to new positions.

Four respondents that offer voluntary testing to new hires told us that "test positives" are hired on a "probationary" basis.

Just over half the firms testing for HIV antibodies (55 percent) began doing so in 1987. Of the others, 33 percent began in 1986, and 7 percent in 1985. Four percent were uncertain of the start date.

A brief overview of individual responses to the AMA questionnaire:

- A hospital tests employees who suffer puncture wounds. Although none have yet tested positive, the organization would refer positives to "appropriate counseling with personal physician," having no AIDS policy at this time.

- A billion-dollar chemical concern offers voluntary testing to selected employees and new hires; testing becomes compulsory if a company physical indicates a necessity. Current employees who test positive are "advised"; they tell us they "might not hire" applicants "if lymphomatic."
- A university offers voluntary testing to current employees and new hires, selecting nursing, medical, and health care students for voluntary testing; counseling is offered to positives.
- A billion-dollar aerospace concern does compulsory "for cause" testing of selected current employees and all new hires. Positives are terminated or not employed.
- A billion-dollar insurance firm offers voluntary testing to selected current employees and new hires at its New York office only. The company reports no action on positives.
- A West Coast fire department offers voluntary testing to selected employees who are exposed to communicable diseases in handling an emergency, and offers counseling and medical assistance to positives.
- A small nonprofit provider of at-home education services offers voluntary testing to all employees and new hires. It requires compulsory testing to ex-convicts up for promotion "due to the prison system and its subcultures." Positives are advised "that they will receive a pay raise but are denied any advancement due to questionable professional standards implied."
- A West Coast police department offers voluntary testing to selected employees "only if work-related contact is proven." Testing is paid for under workers compensation. No action reported for positives.
- A large financial services firm reports that employee participation in blood drives provides voluntary testing and confidential notification of positives; they report no other policy or practice.
- A municipal government agency demands compulsory tests of applicants for emergency medical service jobs. Positives are "not eligible for hire."
- A mid-sized West Coast manufacturer offers voluntary testing to all new hires. No other information.

- A small Midwestern manufacturer offers voluntary testing to all employees and upon promotion; it demands that all new hires be tested and denies employment to positives.

The AMA sample of 995 human resources managers were polled by mail in November 1987. Three hundred ninety-eight respondent firms (40.0 percent) were small companies, defined as having annual sales of less than $50 million; 367 (36.8 percent) were mid-sized ($50 million to $500 million in annual sales); 174 (17.5 percent) were large, with annual sales in excess of $500 million. Fifty-six survey respondents gave no figures for annual sales.

For nonprofit organizations, annual operating budgets are substituted for annual sales.

Four hundred sixty-six respondent firms (46.8 percent) were service providers, 338 (34.0 percent) were manufacturers, and 188 (18.9 percent) were defined as "others," a category that includes diversified conglomerates. Three respondents gave no response to business category.

Appendix B:

How Public Perceptions Impact Business Patronage

The American Management Association commissioned a national opinion poll to measure public attitudes toward AIDS as it relates to the service industry. Louis Harris and Associates conducted telephone interviews with 1,250 people between November 11 and 13, 1987. The results have a ±3 percent margin of error.

Interviewers posed seven hypothetical questions designed to measure how vulnerable certain service providers would be to public knowledge that an employee was infected with HIV. The findings are presented below, along with a brief analysis.

Question 1. Suppose you learned that a waiter at one of your favorite restaurants had a test showing that he had contracted the AIDS virus,* even though he did not have any symptoms of the disease. Do you think you would definitely continue to go to the restaurant, probably continue to go, probably stop going, or definitely stop going?

*The term "AIDS virus" was used rather than "human immunodeficiency virus" to avoid confusion.

Definitely continue	*Probably continue*	*Probably stop*	*Definitely stop*	*Depends*	*Not sure*	*Refused to answer*
129	351	329	381	27	32	1
10%	28%	26%	30%	2%	3%	-

Household income proved an important factor in public attitudes. Among those with an annual household income of over $50,000, 46 percent said they would "probably" or "definitely" continue to patronize the restaurant, while 48 percent said they would probably or definitely stop going there. At the low end, those with an annual household income of $15,000 or less, just 31 percent said they would continue, and 62 percent said they would not.

There were also notable differences according to age and level of education. By age, 47 percent of those under 30 said they would "probably" or "definitely" continue their patronage, compared with just 26 percent of those over 50. Concerning education, fully half of the college graduates said they would "probably" or "definitely" continue, compared with 26 percent of those lacking a high school diploma.

Also, white collar workers (46 percent) were more inclined to continue their patronage than blue collar workers (32 percent).

Question 2. Suppose you learned that a cook at one of your favorite restaurants had a test showing that he had contracted the AIDS virus, even though he did not have any symptoms of the disease. Do you think you would definitely continue to go to the restaurant, probably continue to go, probably stop going, or definitely stop going?

Definitely continue	*Probably continue*	*Probably stop*	*Definitely stop*	*Depends*	*Not sure*	*Refused to answer*
83	228	370	525	17	24	2
7%	18%	30%	42%	1%	2%	-

In every regard, people would be less likely to continue going to a restaurant if a cook were infected with AIDS than if a waiter were so infected. Recall that in the highest income group (over $50,000 per annum), 46 percent said they would "definitely" or "probably" continue to patronize a restaurant where a waiter had the AIDS virus; that number drops to 32 percent if a cook had the virus.

The pattern repeats among all subsets of the sample. The differences by income, age, level of education, and occupation remain, but become more narrow. A regional variation does emerge in responses to this second question, revealing that southerners (77 percent) are more inclined to cease their patronage of a restaurant in such a case than westerners (64 percent).

Question 3. What do you think the owner of a restaurant should do if he learns that one of his waiters has contracted the AIDS virus? Should he fire the waiter, keep him on and tell the customers, keep him on and tell no one, or arrange for him to go on disability?

Fire waiter	*Keep him, tell patrons*	*Keep him, tell no one*	*Put him on disability*	*Depends*	*Not sure*	*Refused to answer*
159	114	197	582	60	129	8
13%	9%	16%	47%	5%	10%	1%

The variety of responses, and the relatively high incidence of interviewees saying "not sure" or "depends," testifies to the complexity of the question. Some respondents may put themselves in the waiter's place, or the owner's; thus the greater inclination to "tell no one" rather than "tell the customers." Still, as *customers,* 60 percent certainly want the waiter off the premises, by disability leave or discharge.

There were few important variations among subgroups. One that did emerge: 23 percent of those lacking a high school diploma said that the waiter should be fired, compared with an average of 11 percent among all others. Also, men (16 percent) were more inclined to see the waiter fired than women (9 percent). Southerners, too, were more in favor of firing the waiter than the rest, by a margin of 5 percent over the national average.

Question 4. What do you think the owner of a restaurant should do if he learns that one of his cooks has contracted the AIDS virus? Should he fire the cook, keep him on and tell the customers, keep him on and tell no one, or arrange for him to go on disability?

Fire cook	*Keep him, tell patrons*	*Keep him, tell no one*	*Put him on disability*	*Depends*	*Not sure*	*Refused to answser*
197	95	161	652	59	78	8
16%	8%	13%	52%	5%	6%	1%

Recall that public response to a cook with the AIDS infection was dramatically different from responses to a waiter so infected (Questions 1 and 2). In contrast, interviewees were nowhere nearly as divided on the issue of what a restaurateur should do if someone on staff had the AIDS virus. Their responses to this query regarding a cook were not importantly different from those regarding a waiter (compare with Question 3).

Again, there were few variations among subgroups, and where they exist they appeared in the same categories as in Question 3. Thirty percent of those lacking a high school diploma thought the cook should be fired, compared with 13 percent of all others (and 10 percent of college graduates). Men (19 percent) were quicker to advise firing the cook than women (13 percent), and those over 65 (21 percent) were more in favor of firing the cook than those under 30 (13 percent).

Question 5. Suppose you learned that a barber or hairdresser that you often go to has had a test showing that he or she has contracted the AIDS virus, even though he or she does not have any symptoms of the disease. Do you think you would definitely continue to go to that barber or hairdresser, probably continue to go, probably stop going, or definitely stop going?

Definitely continue	*Probably continue*	*Probably stop*	*Definitely stop*	*Depends*	*Not sure*	*Refused to answer*
172	340	277	395	19	45	2
14%	27%	22%	32%	2%	4%	-

Barbers and hairdressers have intimate contact with their patrons, and there is at least a hypothetical possibility of a mishap resulting in an exchange of blood—which, unlike food handling and preparation, is a recognized manner of AIDS infection. Nevertheless, the numbers are very similar to those recorded in response to Question 1 (regarding the infected waiter), and not as strong as the reaction to learning that a restaurant's cook was infected.

Once again, income, age, and level of education proved important variables. Thirty-six percent of those with household incomes of $15,000 or less said they would definitely stop going to a hairdresser or barber infected with AIDS, compared with 24 percent of those with incomes of over $50,000. Twice as many interviewees over 65 years of age (46 percent) said they would definitely cease their patronage in such a case as those under 30 (23 percent). Half of those lacking a high school diploma said they would definitely stop going to the barber or hairdresser, compared with 23 percent of college graduates.

Question 6. What do you think the owner of a barbershop or hair salon should do if he/she learns that one of his barbers or hairdressers has contracted the AIDS virus? Should he/she fire the person, keep him/her on and tell the customers, keep him/her on and tell no one, or arrange for disability leave?

Fire person	*Keep person, tell patrons*	*Keep person, tell no one*	*Put on disability*	*Depends*	*Not sure*	*Refused answer*
133	228	242	488	52	99	8
11%	18%	19%	39%	4%	8%	1%

Despite the greater intimacy of the relationship—or perhaps because of it—interviewees are less inclined to want AIDS-infected barbers and hairdressers off the premises than was the case with waiters or cooks (see Questions 3 and 4). Sixty percent of those interviewed thought AIDS-infected waiters should be fired or put on disability, and 68 percent thought the same about cooks, but only half of those interviewed urged one or the other action for barbers and hairdressers. At the same time, 18 percent of those interviewed thought the proprietor of a barbershop or hair salon should tell the customers about AIDS-infected staffers, compared to 9 percent who thought a restaurateur should tell patrons about a waiter so infected.

The education level and gender of the interviewees made a difference in their responses. Twenty-one percent of those lacking a high school diploma thought an HIV-infected hairdresser or barber should be fired, compared with 8 percent of all others. Just 8 percent of women interviewed thought that firing the employee was the proper response, compared with 13 percent of men.

Question 7. Suppose you learned that your dentist also treats some patients who have AIDS. Do you think you would definitely continue to go to that dentist, probably continue, probably stop going, or definitely stop going?

Definitely continue	*Probably continue*	*Probably stop*	*Definitely stop*	*Depends*	*Not sure*	*Refused to answer*
160	292	245	479	40	33	1
13%	23%	20%	38%	3%	3%	-

Although in response to all the questions posed, the number of people who would "definitely" stop their patronage exceeded the number that would "probably" stop, the spread is far more dramatic here than elsewhere—nearly two-to-one. The finding must be chilling to doctors of dentistry. Remember that the matter here is not whether the dentist has AIDS infection, but *whether he or she treats patients so infected.* The public perception conflicts with all that is known and practiced regarding sterilization and disinfection of instruments, and with a 2500-year tradition among medical practitioners to treat all people with disease.

Some variations, none particularly dramatic, appear by subgroup. Only 17 percent of those with incomes above $50,000 say they would definitely continue with a dentist who treats AIDS-infected patients, compared with 11 percent of those with incomes of $15,000 or less. Twenty percent of college graduates would definitely continue, compared with 10 percent of all others; 16 percent of white collar workers would continue, compared with 9 percent of blue collar workers.

Appendix C:

How to Run an Effective Educational Program

EDITOR'S NOTE

Material in this appendix has been excerpted from *Facilitating AIDS Education in the Work Environment,* a 64-page manual produced by the Orange County Business Leadership Task Force on AIDS and Alcohol and Drug Abuse. Other chapters in the publication present facts about AIDS, answers to questions commonly asked when implementing an education program, listings of community resources, descriptions of educational materials, and sample corporate policies. Pacific Mutual, the flagship company of the Pacific Financial Companies, coordinated and staffed development. Chapter 1 is reproduced herein through special permission from Pacific Mutual. (Copyright 1987. All rights reserved.) Readers wishing to obtain a copy of the complete publication should contact Pacific Mutual Life Insurance Company, 700 New Port Center Drive, New Port Beach, California, 92660.

To be effective in educating your employees, you need to make AIDS information readily available to them. Information should take a variety of forms: workshops, brochures, posters, newsletter articles. People are different and respond to different types of education. One approach may not be effective in reaching everyone. A variety of approaches will ensure that the greatest number of people will be reached by your AIDS education programs.

You may already have an effective method of educating employees. If so, use it to educate them about AIDS. If, however, you want to enhance an already existing program, or if you are creating a new program specifically for AIDS education, the following paradigms will be useful. You can implement them individually if you choose, or as a complete AIDS education program.

Small businesses may be able to implement only one or two of the programs described in this section. For instance, it may not be feasible for a company of 50 employees to take one or two hours out of the workday to conduct a workshop. In that case, it probably would be easy to display posters and to distribute brochures and a letter from management.

In either case, any amount of information that is distributed is useful in combatting this disease and in saving lives.

Remember that repetition is a key element of this program. Changing behaviors and reducing fear are not easily accomplished—and certainly not without repeated emphasis on the basic messages.

RESOURCE/EDUCATION COMMITTEE

Large companies may want to establish an education committee with representatives from several departments including personnel, medical, human resources, employee communications, employee assistance, food service, and maintenance. Once selected, these representatives need to be instructed about the issues surrounding AIDS, the company's policies and the availability of community resources. Local agencies (listings for California can be found in the "Community Resources" section) can assist in training the committee.

Once trained, this committee can then work together to design the company's educational program and respond to employee questions and concerns. Members of the committee can serve as contacts for employees to get further information or community referrals. It is useful to have a variety of departments represented on this committee so that the special concerns of all the employees are met.

While it would be beneficial to use the committee process to develop your program, any individual appointed to organize an employee education program can easily handle the task with the assistance of this manual and the utilization of local resources.

EMPLOYEE MEETING AND WORKSHOPS

One effective method of providing education to employees is through group employee meetings or in-service training sessions. They are relatively easy to conduct and are effective in reaching employees. Members of the resource committee or other knowledgeable employees could conduct the meeting. If you do not have resources within the company, many of the agencies listed in the "Community Resources" section of this manual, or your local department of public health, can provide information about speakers who are available for these types of meetings. In fact, the American Red Cross offers a workshop entitled "Working Beyond Fear" dealing specifically with AIDS education in the workplace (see "Community Resources" for details).

If you choose to conduct the program yourself, follow these basic steps in planning an effective employee meeting or workshop:

When:

There are several alternatives as to when to conduct an AIDS workshop. It can be incorporated into an existing employee seminar series. It can be part of a department update session, staff meeting or even a specially-planned program. The meeting or meetings can even be conducted during the lunch hour. Providing time during the workday for employees to attend, and strongly encouraging them to do so, is preferable.

Who:

All employees should be educated. AIDS is a deadly disease, and while there is no risk of contracting it through casual contact in the workplace, employees do need to be told how to prevent transmission of the virus outside the workplace if education programs are going to be effective in helping to stop the transmission of the AIDS virus.

Begin by educating top management. They can then assist in motivating middle management to become involved in AIDS education. Then, the remainder of the staff should be educated. Each manager, with the support of the education committee, will then be able to take some responsibility for educating his or her staff. If your company has a housekeeping staff, be sure they are also included in the educational program. These employees are often exposed to a host of bodily fluids.

Format:

Keep it simple. You want a program that is easy to implement and takes relatively little support.

Again, if you have adequately trained your education committee members, utilize them in facilitating these presentations. And, many of the agencies listed in the "Community Resources" section of this manual or your local department of public health will be able to provide information on speakers for your program. The presence of a local expert as a resource person is highly recommended.

Introduction (5 - 10 minutes)

Begin by welcoming employees to the program and thank them for taking the time to learn the facts about this deadly disease. If your company has a personnel policy regarding AIDS or catastrophic illness, state it.

Stress that AIDS is not casually transmitted, particularly in the workplace. Tell them this program will explain why this is the case. Emphasize that you want to answer their questions and that there are no dumb questions when dealing with a disease such as AIDS.

Introduce any videotapes you might be showing. Tell the

audience that the video provides the basic facts about AIDS and that a discussion will follow.

Videotape (15 - 30 minutes)

There are a variety of videotapes available for use during the workshop. See the "Educational Materials" section of this manual for descriptions, ordering and rental information. Pay particular attention to the available descriptions of these videos. Some are designed specifically for teenagers, others for an older audience. Be sure to preview any video before you use it to ensure that it is appropriate for your particular employees.

Discussion (30 minutes)

Follow the videotape with a brief discussion about the information presented in the videotape. After doing so, ask for questions from the audience (see the "Common Questions About AIDS" section of this manual for answers to the most frequently-asked questions). An outside resource person can be most useful in answering employees' questions.

Closing Remarks and Brochure Distribution (5 minutes)

As the program concludes, distribute general information brochures to the audience (sample brochures and ordering information are included in the "Educational Materials" section of this manual). Let the audience know who in the company can provide them with further information. Also, let the employees know about local community resources that are available. You may even want to distribute appropriate portions of the "Community Resources" section of this manual to the group.

Some people may be embarrassed about asking questions. Prepare and distribute a workshop evaluation form to the audience, and be sure to provide space on the form for these questions. Answer them later in a follow-up mailing or employee newsletter. (See pages 155-160 for a sample.)

Close by again stressing the importance of AIDS education. Request that people take the information they have learned and share it with family and friends.

Below is an outline of a typical workshop format as discussed above. You may want to have it with you when conducting a workshop.

AIDS WORKSHOP OUTLINE

I. Introduction (5–10 minutes)
   A. Welcome
   B. Personnel policy (if applicable)
   C. Introduce videotape

II. Videotape (15–30 minutes depending on video selected)

III. Discussion (30 minutes)
   A. Highlight important facts discussed in the videotape
   B. Respond to questions from the group (have "Common Questions About AIDS" section with you)

IV. Closing Remarks and Brochure Distribution (5 minutes)
   A. Distribute AIDS information brochures and workshop evaluation forms
   B. Highlight company and community resources
   C. Stress education and sharing information with family and friends

BROCHURE AND POSTER DISTRIBUTION

Another effective method of reaching employees is through the distribution of brochures. Described in this manual are a variety of educational brochures useful for distribution in the workplace. Ordering information is contained in the "Educational Materials" section of this manual. There are a couple of brochures specific to the workplace that would be appropriate to send to every employee, either through the mail, with paychecks or via your current method of distribution. Other brochures geared to specific groups such as homosexuals, or children, should be made available to employees in a manner that ensures privacy.

Posters on company bulletin boards are useful in providing visual reminders that AIDS is a part of our society and we have to deal with it. They also serve to remind employees that the company is concerned about AIDS and wants its employees to be concerned as well. The American Red Cross and the San Francisco AIDS Foundation both have posters available.

VIDEO LENDING LIBRARY

The "Educational Materials" section of this manual contains information about a variety of educational videotapes. Some are geared for the workplace, others for teenagers (suitable for use in the home or with young employees). Establishing an AIDS video lending library where employees can check out videos for use at home is an effective way of communicating the information to the families and friends of your employees.

SAMPLE NEWSLETTER ARTICLES AND LETTERS FROM MANAGEMENT

Yet another useful method of spreading information about AIDS is through employees newsletter articles. In addition, a letter from the CEO to every employee is useful in conveying information about AIDS. The following pages contain samples of both. It is useful to distribute a listing of services and materials available to employees from the company. This will further help to ensure that everyone knows information is available.

SAMPLE EMPLOYEE LETTER ON AIDS

TO: All Employees

FROM: Chief Executive Officer

SUBJECT: AIDS AWARENESS AND EDUCATION PROGRAM

AIDS, Acquired Immune Deficiency Syndrome, is an infectious disease which kills thousands of people annually. In 1981, when AIDS was first identified, it was thought to be limited to homosexuals and drug abusers. Now in 1987, it is clear that heterosexuals and homosexuals alike can contract AIDS. The disease continues to spread, and there is still no cure in sight. The danger of AIDS to society is so great that we feel you cannot afford to be unaware of the facts. We want to make you aware of how AIDS is contracted and how it is not—and how to prevent infection from the virus.

With this in mind, the company is introducing an AIDS Awareness and Education Program for all employees. The program will include educational seminars with health experts to discuss the most current information about AIDS and to answer your questions. In addition, we will make videotapes available on a "lending library" basis to bring the educational aspects of the program to your families. Articles in the monthly newsletter will discuss the latest information on AIDS. AIDS awareness material will be placed on bulletin boards throughout the company. Educational brochures will be made available for you and your family.

One goal of the AIDS Awareness and Education Program is to dispel myths about AIDS. For example, there is a popular misconception that AIDS can be contracted through casual contact with a person carrying the AIDS virus. This is not true. Sharing workspace or telephones with a person who has AIDS does not increase your risk of getting AIDS. This program intends to rid individuals of misconceptions and unnecessary anxiety about AIDS.

We want to help you become informed through the company's AIDS Awareness and Education Program. Our hope is that you will share with your family what you learn about AIDS. It is only through awareness that we can prevent the spread of this killer disease.

Sincerely,

Chief Executive Officer

SAMPLE EMPLOYEE NEWSLETTER ARTICLE #1 ON AIDS

The discovery of any new disease is a natural cause for public concern. The identification of Acquired Immune Deficiency Syndrome (AIDS) in 1981 was no exception. What makes AIDS unique is that fear is growing among those considered at low risk, despite scientific assurance that it is not an easy disease to catch.

If you are one of the many who are at low risk of contracting AIDS, but who are nevertheless worried about it, it is important to arm yourself with a good base of information. AIDS is a fatal disease for which there is currently neither a cure nor a vaccine. This being the case, public education on the subject of AIDS is of paramount importance. Education programs in the workplace aimed at dispelling fear and misinformation about AIDS encourage employees to learn the facts. Misconceptions about AIDS abound. For example, contrary to the popular casual-contact myth, the AIDS virus is not spread through normal daily contact at work, in school, or at home.

Here, at our company, we have introduced a comprehensive worksite education program about AIDS. We hope you will make participation in this program a personal priority. It is important for all of us to recognize that we can all join the battle against AIDS by fighting fear with facts.

SAMPLE EMPLOYEE NEWSLETTER ARTICLE #2 ON AIDS

FACTS ABOUT ACQUIRED IMMUNE DEFICIENCY SYNDROME*

AIDS, or Acquired Immune Deficiency Syndrome, is a deadly and frightening disease which was first identified in 1981. To date, there have been more than 35,000 cases of AIDS reported in the United States. More than half of these people have died. The cause has been established as a specific virus known as HIV. The virus attacks the

*An article by David B. Dunkle, M.D., reprinted from the Hershey Foods company newsletter.

body's immune system making it vulnerable to rare opportunistic infections and cancers which, in turn, result in fatalities. There is no cure yet, but research is in progress. There is a specific blood test which can tell when an individual has been exposed to the HIV virus.

AIDS-Induced Panic Syndrome (AIPS) is a term used to refer to the almost frightening behavior resulting from the vast amount of misinformation and misunderstanding about AIDS. A number of children have been refused admission to school, families of AIDS patients have refused to care for them, and even some ambulance companies have refused to transport people with AIDS. Although there is no cure yet for AIDS, education is the cure [for the spread] of AIDS. Much of the irrational behavior is based on poor information regarding how AIDS is spread.

AIDS is difficult to catch. The virus is not easily spread from person to person, and has never been spread by casual contact. Medical authorities who have studied thousands of cases agree that AIDS is contracted in these four ways:

- Sexual intercourse with a person infected with HIV virus (both homosexual and heterosexual intercourse);
- Sharing intravenous drug needles or syringes with an infected person;
- Injection of contaminated blood products such as blood transfusions. This method of transmission should no longer be as likely to occur because blood is now being screened for contamination;
- A woman infected with AIDS virus who becomes pregnant can pass it to her baby.

The AIDS virus cannot be passed through the air. Sneezing, breathing or coughing do not spread AIDS. AIDS is not transmitted through preparation or serving of food or beverages. No cases of AIDS have resulted from casual contact.

Even when children have played, eaten, slept, kissed and used the same toothbrush with a brother, sister or parent suffering from AIDS, none has become infected. No cases have ever been linked to sharing typewriters, telephones, tools, papers, water fountains, chewed pencils, eating facilities, showers, or even toilet seats.

Our company has not yet had to deal directly with a person with AIDS, but if and when it happens it is our hope that our employees will be well informed, act rationally and give the same support and concern they would like to get themselves if they had a serious disease.

SAMPLE EMPLOYEE MEMO ON AIDS AND THE WORKPLACE*

TLC is committed to providing a work environment which allows all of us to perform our jobs in a safe and productive manner. Our Business Philosophy, which states that we respect the dignity and worth of every person, reinforces this commitment, as does our Equal Employment Opportunity statement, which explains our policy and practice with respect to prohibiting discrimination in every phase of employment

Employees with a handicap or medical condition (life threatening or otherwise), who are able to meet appropriate performance standards and whose continued employment does not pose a threat to their own health and safety or that of others, are assured equal employment opportunities and reasonable accommodation in their employment here. This has been and continues to be the policy and practice of the Transamerica Life Companies.

I am reemphasizing this policy because of the media attention currently being given to Acquired Immune Deficiency Syndrome (AIDS) and its impact on the workplace. Any employee who contracts AIDS will be guaranteed the same rights as any other employee who has a handicap or illness.

The "AIDS issue" is surrounded by myths and misunderstandings. To deal with it openly and honestly in the workplace, we must begin with a factual foundation. Over the next couple of months, we will introduce special educational programs providing factual,

*A memo to all employees from David R. Carpenter, Chairman and CEO, Transamerica Life Companies, February 20, 1986.

medical information on AIDS, how it can be transmitted, and more importantly, how it cannot be transmitted. This should help to dispel rumors, lessen fears and raise our comfort level about this dreaded disease.

If you have any questions or need assistance at any time, please contact either EEO Administration on x or Employee Relations on x

OTHER SUGGESTIONS

Buttons

One way to increase employee awareness and responsibility about AIDS is to provide each employee with a button that indicates he or she participated in an AIDS workshop. This will create an atmosphere of awareness and responsibility as more and more employees are educated and begin wearing the buttons. The buttons should generate further discussion about AIDS among employees beyond the bounds of the workshop.

Special Workshops for Managers

Managers face many special challenges in dealing with an employee with AIDS. A manager needs to be prepared to deal with confidential information. Further, he or she needs to be prepared to deal with the special needs of the person with AIDS. The manager must also be prepared to address the concerns of other employees in his or her unit.

It may be useful to hold a "For Managers Only" session on AIDS to discuss the special issues that a manager will face when confronted with an employee or employees with AIDS. Perhaps a panel with a personnel representative, medical and legal counsel, and an AIDS educator could provide the necessary additional information that managers need to deal with the issues surrounding an employee with AIDS.

Special Programs for Parents

In order for education to be effective, it must be shared with others. Many parents are trying to decide how to educate their young children and teenagers. But many of them do not know where to begin. A special program for parents discussing how to present AIDS information to children could help many parents in dealing with this issue. If you set up a video lending library, you may want to host lunch hour preview screenings of the videos produced for children and teenagers. This will allow parents to select from among the available videos before taking them home to view with children.

Sexually Explicit Material

Because AIDS is primarily a sexually-transmitted disease, many issues surrounding human sexuality, including many homosexual acts, often need to be addressed. Not all employees may want or need this information; however, others will and do need it. Specific information about safer sexual practices is critical for those employees who are sexually active.

One possible way to distribute this information to employees without offending anyone is to do a mailing of the information to all employees. This should be done using a sealed envelope, and on the envelope something similar to the following could be printed: "The information contained in this packet discusses sexual material surrounding AIDS risk reduction. It is meant primarily for people in one of the high-risk categories: sexually-active homosexual or bisexual men, people with multiple sexual partners (male or female), sexual partners of people with AIDS or a positive antibody status (male or female), and sexual partners of IV drug abusers. If you are offended by, or have no need for this information, pass it along to someone else or dispose of it."

Business and Civic Association Groups

Groups such as chambers of commerce, industrial leagues, merchant associations, and shopping mall management groups are in an ideal position to provide AIDS education to their members.

They have the opportunity to reach a large number of businesses with one coordinated and centralized effort. Many of the educational programs discussed in this section can be easily adapted for use by these associations. Workshops, brochure and informational letter mailings, and video lending libraries all would be of great benefit to members of these groups. Such programs would be especially useful in reaching many people who are employees of small businesses.

Program Evaluation

You should incorporate a program evaluation form into your educational programs. This form will allow you to assess the knowledge level of your employees as well as the effectiveness of your educational programs. The evaluations should be anonymous so that employees will feel comfortable responding to the questions. A sample evaluation from Wells Fargo Bank follows:

WELLS FARGO BANK, N.A.
EMPLOYEE ASSISTANCE SERVICES

AIDS EDUCATION PROGRAM
CONFIDENTIAL EVALUATION FORM

Please do not write your name on this form as *your answers are confidential.* We would like your help in evaluating the AIDS Education Program by completing this confidential evaluation form. Your opinion will help us determine how we can best offer this information to other employees.

A. Please circle the number that applies to you:

1. Are you:	2. Your Race: (optional)	3. Your age group:
1 Female	1 American Indian	1 18-24 yrs
2 Male	2 Hispanic	2 25-30 yrs
	3 Black	3 31-37 yrs
	4 White	4 38-45 yrs

Are you:	Your Race: (optional)	Your age group:
	5 Asian	5 46-55 yrs
	6 Philippine	6 56-over
	7 Other	

B. Using the sentence and scale below, please circle the number that best describes your reaction to the AIDS Education Program:

Before you received information *at work* about AIDS. . .

	Not At All	*Not Too Much*	*Somewhat*	*Moderately*	*Very Much*
1. Were you concerned about contracting (catching) AIDS?	1	2	3	4	5
2. Did you know how AIDS was transmitted?	1	2	3	4	5
3. Were you concerned about contracting AIDS from an employee through normal work contact?	1	2	3	4	5
4. Did you know how to prevent contracting the AIDS virus?	1	2	3	4	5

	Not At All	Not Too Much	Somewhat	Moderately	Very Much
5. Did you change your behavior to minimize your risk of contracting the AIDS virus?	1	2	3	4	5

C. Using the sentences and scale below, please circle the number that best describes your reaction to the AIDS Education Program.

Did you find that the information. . . .

	Not At All	Not Too Much	Somewhat	Moderately	Very Much
1. Was new to you?	1	2	3	4	5
2. In the brochures was helpful to you?	1	2	3	4	5
3. In the memo was useful to you?	1	2	3	4	5

D. *After* you received information *at work* about AIDS. . .

1. Are you now concerned about contracting AIDS?	1	2	3	4	5
2. Do you now feel you know how AIDS is transmitted?	1	2	3	4	5

	Not At All	*Not Too Much*	*Somewhat*	*Moderately*	*Very Much*
3. Are you now concerned about contracting AIDS from an employee through normal work contact?	1	2	3	4	5
4. Do you now know how to prevent contracting AIDS?	1	2	3	4	5
5. Do you now plan to change your behavior to minimize your risk of contracting AIDS?	1	2	3	4	5

E. Circle the number that best describes your opinion.

	No Value At All	*Of Little Value*	*Partially Helpful*	*Moderately Valuable*	*Totally of Value*
1. Overall, how would you rate the value of the AIDS Education Program?	1	2	3	4	5

	No Value At All	*Of Little Value*	*Partially Helpful*	*Moderately Valuable*	*Totally of Value*
2. How would you rate the value of watching a video to have your questions answered?	1	2	3	4	5
3. How valuable was the question/answer period with the medical expert? (if applicable)	1	2	3	4	5

F.

1. Were all of your questions addressed and answered by the video presentation?

Yes No Partially (check one)

2. Do you have any further questions that you would like to have answered?

____

____

____

____

3. Do you feel you need/want additional education about AIDS:

Yes No (check one)

4. Do you have any other comments/suggestions?

____

____

____

____

Thank you for taking the time to complete this evaluation form.

ONE OF OUR OWN

The American Management Association is distributing the video ONE OF OUR OWN along with Dartnell Corporation (see page 82 of this manual for a review of the film). AMA will donate all profits earned from distribution of the video to AIDS research.

If you wish to order through the American Management Association, call or write:

AMA Film/Video
9 Galen Street
Watertown, Massachusetts 02172
800-225-3215 (outside Massachusetts)
617-926-4600 (in Massachusetts)

Prices (nonmembership and AMA membership, respectively):
$45/$40 to preview
$195/$185 to rent
$565/$535 to buy

AMA Membership Publications Division
American Management Association
135 West 50th Street, New York, N.Y. 10020
2333